AQA Geography
Revision Guide

A LEVEL & AS

PHYSICAL GEOGRAPHY

Series editor Tim Bayliss

Lawrence Collins

Alice Griffiths

Simon Ross

OXFORD
UNIVERSITY PRESS

OXFORD
UNIVERSITY PRESS

Great Clarendon Street, Oxford, OX2 6DP, United Kingdom

Oxford University Press is a department of the University of Oxford.
It furthers the University's objective of excellence in research, scholarship,
and education by publishing worldwide. Oxford is a registered trade mark of
Oxford University Press in the UK and in certain other countries

© Oxford University Press 2018

Series editor: Tim Bayliss

Authors: Lawrence Collins, Alice Griffiths, Simon Ross

British Library Cataloguing in Publication Data

Data available

ISBN 978-019-843266-1

10 9 8 7 6 5 4 3 2 1

Paper used in the production of this book is a natural, recyclable product made
from wood grown in sustainable forests. The manufacturing process conforms to
the environmental regulations of the country of origin.

Printed in Great Britain by Bell and Bain Ltd., Glasgow

Acknowledgements

The publisher and authors would like to thank the following for permission to
use photographs and other copyright material:

Cover: David Baker/Trevillion Images; Earth Imaging/Getty Images; **p8(t):**
Ralph Hagen/Cartoonstock; **p8(b):** Randy Glasbergen; **p18:** PJ photography/
Shutterstock; **p25:** Citizen of the Planet/Alamy Stock Photo; **p29:** courtesy
of NOAA; **p36:** Nigel Dickinson/Alamy Stock Photo; **p39:** South West Water;
p46: David South/Alamy Stock Photo; **p47:** Simon Ross; **p48:** Sylvia Kania/
Shutterstock; **p51:** imageBROKER/Alamy Stock Photo; **p53:** VT750/Shutterstock;
p58, 63: Simon Ross; **p68:** John McLellan/REX/Shutterstock; **p70:** Tim Bayliss;
p71: imageBROKER/Alamy Stock Photo; **p72:** Geography Photos/Getty Images;
p76: GEORGE BERNARD/SCIENCE PHOTO LIBRARY; **p77:** NASA; **p80:** Matt
Adamson/Shutterstock; **p81(t):** geogphotos/Alamy Stock Photo; **p81(b):**
A.P.S. (UK)/Alamy Stock Photo; **p82:** odisha_factsheet-1; **p83:** Uncredited/AP/
REX/Shutterstock; **p86:** Simon Ross; **p95:** Photograph courtesy from www.
markhewittphotography; **p96:** Jo Katanigra/Alamy Stock Photo; **p97:** funkyfood
London - Paul Williams/Alamy Stock Photo; **p98(t):** david speight/Alamy Stock
Photo; **p98(b):** DR. MARLI MILLER/VISUALS UNLIMITED, INC. /SCIENCE PHOTO
LIBRARY; **p99:** Design Pics Inc/Alamy Stock Photo; **p100:** Tom Bean/Alamy
Stock Photo; **p103:** Dmitry Lovetsky/AP/REX/Shutterstock; **p104:** Mark Thorpe/
Creative Commons Attribution-Share Alike 3.0 Unported; **p105:** Image/photo
courtesy of the National Snow and Ice Data Center, University of Colorado,
Boulder; **p107:** Erlend Bjørtvedt (CC-BY-SA); **p108:** with kind permission from
Svalbard Expeditions; **p111:** Barcroft/Getty Images; **p118:** S-F/Shutterstock;
p119: club4traveler/Shutterstock; **p120:** NASA Earth Observatory; **p121:** Wead/
Shutterstock; **p125:** STR/Getty Images; **p126:** REX/Shutterstock; **p127:** G Allen
Penton/Shutterstock; **p128:** The Asahi Shimbun/Getty Images; **p129:** Daniel
Berehulak/Getty Images; **p130:** Jeff Schmaltz, MODIS Rapid Response Team,
NASA/ GSFC; **p133:** Simon Ross; **p136:** NASA/EARTH OBSERVATORY/JOSHUA ST;
p143: dcwcreations/Shutterstock; **p145:** Aleksey Stemmer/Shutterstock; **p151:**
Travel Ink/Getty Images; **p156(t):** PJ photography/Shutterstock; **p156(b):** Rebecca
Cole/Alamy Stock Photo; **p159:** keith burdett/Alamy Stock Photo; **p160(l):**
Avalon/Construction Photography/Alamy Stock Photo; **p160(r):** Tim Cuff/Alamy
Stock Photo; **p162:** South West Water; **p163:** Keith Bath/Shutterstock.

Artwork by Kamae Design, Aptara Inc., and Giorgio Bacchin (p27(b), 32).

Every effort has been made to contact copyright holders of material reproduced
in this book. Any omissions will be rectified in subsequent printings if notice is
given to the publisher.

Contents

Contents

Contents

Introduction: a guide to success

Your revision guide

This revision guide contains key revision points that you need to prepare for Paper 1 of the AQA GCE Geography specification. It meets both the content requirements of the A Level course (Physical geography), but can equally well be used for the separate AS course (Physical geography and people and the environment).

Both this and its sister *Human Geography Revision Guide* need to be used alongside Oxford's AQA Physical and Human Geography student books. The revision guide provides links to the student books where necessary.

You will also find helpful other resources in this series including Oxford's *AQA Geography A Level and AS Exam Practice and Skills* book and *Kerboodle* digital resources and assessment online.

This revision guide provides:

- targeted AS and A Level revision guidance including top tips for exam success
- easy to digest spec-specific content, recaps and summaries
- flexible revision checklists to help you monitor your progress.

Key content of each section in the student book is summarised in a double or single page (Figure **1**).

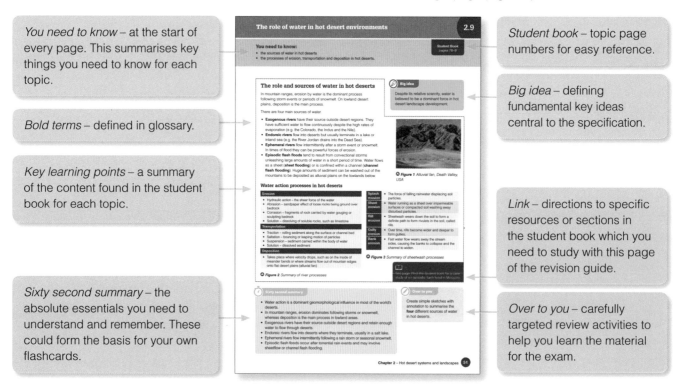

You need to know – at the start of every page. This summarises key things you need to know for each topic.

Bold terms – defined in glossary.

Key learning points – a summary of the content found in the student book for each topic.

Sixty second summary – the absolute essentials you need to understand and remember. These could form the basis for your own flashcards.

Student book – topic page numbers for easy reference.

Big idea – defining fundamental key ideas central to the specification.

Link – directions to specific resources or sections in the student book which you need to study with this page of the revision guide.

Over to you – carefully targeted review activities to help you learn the material for the exam.

⊙ Figure 1 *Your revision guide's key features*

Guided answers to the activities can be found at www.oxfordsecondary.co.uk/geography-answers

In addition, each chapter introduction contains information on how the content relates to both AS and A Level specifications. You are also encouraged to build a record of essential key terms, and **either** track your revision progress **or** use the guidelines to indicate topics you are more or less confident about.

What does the AS specification include?

The AS Geography qualification does not count towards an A Level – it is a qualification in its own right. The specification consists of six topics, of which you must study three plus a fieldwork investigation. They are grouped into two papers: *Physical geography and people and the environment; Human geography and geography fieldwork investigation.*

Paper 1: Physical geography and people and the environment	Paper 2: Human geography and geography fieldwork investigation
• Section A: you must study **one** of *Water and carbon cycles* **or** *Coastal systems and landscapes* **or** *Glacial systems and landscapes*. • Section B: you must study **either** *Hazards* **or** *Contemporary urban environments*.	• Section A: *Changing places*. **This topic is compulsory**. • Section B: *Geography fieldwork investigation and geographical skills*. You will have to answer one question on each of these topics.

✎ Over to you

Complete this for Paper 1: Physical geography and people and the environment

My optional topic for Section A is

....

.... (Chapter ....).

My optional topic for Section B is

....

.... (Chapter ....).

Remember, there is no coursework at AS Level, but questions will be included in Paper 2 about the fieldwork you have completed as part of your AS course.

What does the A Level specification include?

The A Level specification consists of eleven topics of which you must study six plus a fieldwork investigation. They are grouped into three papers: *Physical geography; Human geography; Geography fieldwork investigation.*

Paper 1: Physical geography	Paper 2: Human geography	Paper 3: Geography fieldwork investigation
• Section A: *Water and carbon cycles*. **This topic is compulsory**. • Section B: you must study **one** of *Hot desert systems and landscapes* **or** *Coastal systems and landscapes* **or** *Glacial systems and landscapes*. • Section C: you must study **either** *Hazards* **or** *Ecosystems under stress*.	• Section A: *Global systems and global governance*. **This topic is compulsory**. • Section B: *Changing places*. **This topic is compulsory**. • Section C: you must study **either** *Contemporary urban environments* **or** *Population and the environment* **or** *Resource security*.	• You must complete an individual investigation which must include data collected in the field. • Your investigation must be based on a question or issue defined and developed by you relating to any part of the specification content.

✎ Over to you

Complete this for Paper 1: Physical geography

My optional topic for Section B is

....

.... (Chapter ....).

My optional topic for Section C is

....

.... (Chapter ....).

How to revise productively

Think about this introduction's opening comments on the challenge of AS and A Level. Revision doesn't have to be a bore or even a chore. With the right mindset it can be enormously satisfying, not least in the latter stages when all the pieces of the jigsaw start to come together, and you're anticipating what's coming next.

Rather like the 'solution' to predicting earthquakes, you and your friends' GCSE revision experiences should have convinced you that there is no 'one size fits all' solution to effective learning, memorising and/or revising. However, you'll know by now that revision has to be:

- scheduled, planned and **organised**
- **active** if it is going to be productive
- done **without distractions**
- in tune with your body clock and concentration patterns.

You are strongly encouraged to embrace the general 'good practice' explained by your teachers and the tips provided in this guide.

Use revision techniques that work for you, but don't avoid trying new ideas. You never stop 'learning to learn', and a new learning, memorising or revision aid tried now may prove to be invaluable – not just in these examinations, but for those at a higher level in years to come.

Think back to your GCSE revision. Reflect on how successfully (or not!) you:

- applied advice and guidance on planning your revision
- kept to and adapted your revision timetable if and when necessary
- experienced and understood your concentration curve (Figure **4**)
- managed your working environment, stress and distractions
- used different techniques
- practised using past exam questions.

Positive reflection, with the intention of learning from both your good and bad experiences of revision will pay dividends. So:

- consider the range of techniques you could use
- adopt what you know to be valuable
- be prepared to adapt and change.

I REALLY CRAMMED LAST NIGHT.

🔼 *Figure 2*

CartoonStock.com

"It's called 'reading'. It's how people install new software into their brains"

🔼 *Figure 3*

Remind yourself of revision techniques, and add your own in the spaces provided. Tick the ones that work for you or add your own alternatives.

Mind-mapping (spider diagrams)	
Mnemonics	
Flashcards	
Topic posters	
Factfiles	
Lists of key terms/words	

Whatever 'old favourites' worked for you – use them again. Just because you were revising for GCSE doesn't devalue them at AS or A Level. But also consider taking your revision to the next level – there might be something here that transforms the way you work (Figure **5**).

Finally, you've probably already tried revising with a partner or in small groups. Message, Skype and Facetime interaction can be productive for some, but a persisting distraction if not scheduled and executed with real discipline.

Revising in class also needs self-discipline, but you should never underestimate the value of extra time spent with your teachers discussing topics you're not clear about, looking at exam questions and understanding mark schemes. However, it is revising alone that will dominate at this level – so make it count!

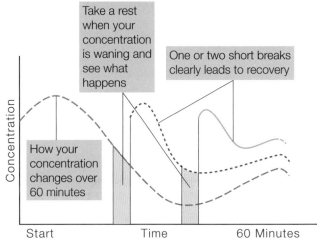

Figure 4 *How short rests boost concentration*

Fixed word count abstracts
Set yourself a fixed word count (say 50 or 100) for a topic summary. Once drafted, 'pruning' to the exact word count forces your mind to concentrate on identifying key information. Writing out the finished summary fixes knowledge into your long-term memory, so is more easily recalled when next revised.

Beware the highlighter pen!
Never highlight this guide or your notes straight away:
- 'Dribble' in soft erasable pencil in the early revision stages.
- Only use the ink highlighter when you're absolutely certain.
- Only use a colour you like.

Revising key terms and command words
You'll have revision sessions when it just isn't going in, so change tack. For example:
- Focus on updating your revision progress and compiling the key terms in your chapter introductions (make use of the glossary).
- Remember – one of the biggest exam pitfalls is misunderstanding command words. So check and revise them.

Dealing with distractions
- Be an active learner – focus on the specific task, not the specified time.
- Leave your phone in another room – you can check it during your regular breaks.
- Divide each study session into short-range goals which demand your full attention.

Taking your revision to the next level

Improving long-term memory by repetition
66% of material is lost within seven days if not reviewed again – and 88% is gone after six weeks! So build *review* time into your daily and weekly revision timetable. It will save you having to re-learn material from scratch!

Reading better and faster – the PQ2R method
P = Preview Begin with a quick skim to get an overview. Look for section headings, figure captions and key words.
Q = Question Look for answers to 'What?', 'Why?', 'Where?', 'When?' and 'Who?' questions which will identify the main learning point(s). This is active learning.
R = Read Now read the spread carefully, with these questions in mind. Your mind will actively look for answers. Make brief summary notes. Slow down over difficult sections.
R = Review Check your understanding by reviewing and testing your recall. Check your notes and answer the initial questions.

Practising output
Remember – exam preparation involves giving out information, not just taking it in! So:
- work on past exam papers
- follow the marks – see how marks are allocated by using the mark schemes.
- try a complete exam paper against the clock.

This will help you to avoid major surprises come the 'real thing'.

Figure 5 *Taking your revision to the next level*

Top tips for exam success

Success in your AS or A Level exam will involve three ingredients:

- having a thorough knowledge and understanding of the subject matter
- making the most of this knowledge and understanding in the exam through effective exam strategy and technique
- managing stress (Figure **6**).

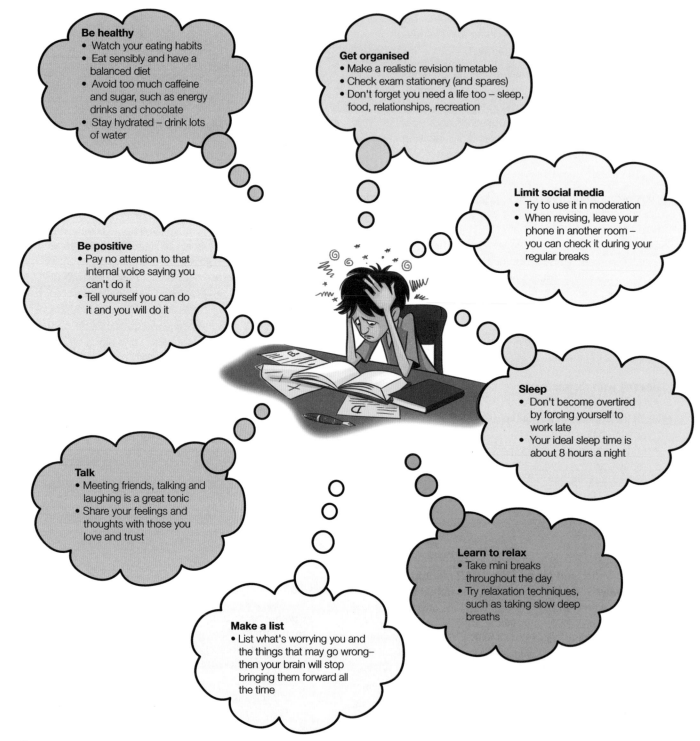

Be healthy
- Watch your eating habits
- Eat sensibly and have a balanced diet
- Avoid too much caffeine and sugar, such as energy drinks and chocolate
- Stay hydrated – drink lots of water

Get organised
- Make a realistic revision timetable
- Check exam stationery (and spares)
- Don't forget you need a life too – sleep, food, relationships, recreation

Limit social media
- Try to use it in moderation
- When revising, leave your phone in another room – you can check it during your regular breaks

Be positive
- Pay no attention to that internal voice saying you can't do it
- Tell yourself you can do it and you will do it

Sleep
- Don't become overtired by forcing yourself to work late
- Your ideal sleep time is about 8 hours a night

Talk
- Meeting friends, talking and laughing is a great tonic
- Share your feelings and thoughts with those you love and trust

Learn to relax
- Take mini breaks throughout the day
- Try relaxation techniques, such as taking slow deep breaths

Make a list
- List what's worrying you and the things that may go wrong– then your brain will stop bringing them forward all the time

⬥ **Figure 6** *Managing stress*

Introduction: a guide to success

Your golden rules

Students who perform well almost always follow these rules.

They **revise** – thoroughly and productively (pages 8–9).

They know which **topics** will be in each exam (page 12).

They **practise** answers under timed conditions, including whole papers.

They are familiar with **mark schemes** and know what examiners are looking for to reach the highest **levels** (pages 13–14).

They **read the front cover instructions** – calming themselves by confirming what they expect.

They display and check an **exam timetable** with exact details of what, where and when (paper, venue, day and time).

They **read the paper carefully** – allowing time to 'size-up' and dissect questions before starting (Figure **7**).

'Outline' is the command word, so provide a brief account of relevant information

Relate everything you say to weathering processes, and why they happen in the way they do

Outline the impact of temperature variation on weathering processes in hot deserts. **[4 marks]**

The focus of the question

This is a 4-mark question, so it must be point marked (correct points and development)

⬥ **Figure 7** *Dissecting the question*

They understand and respond correctly to **command** words (page 15).

They look at the **marks**, which indicate which questions will be point marked and which level marked (page 13 and Figure **7**).

They **answer every question required** and leave no blanks.

They write in **full sentences** using appropriate, **specialist vocabulary**.

They apply **specific located details** about case studies or examples.

They work out in advance and practise how long to spend on questions.

They never 'steal' extra time on favourite topics to the detriment of others.

They get the **timing** right – especially on longer answer questions – including 5–10% planning time, 80–85% writing time and 10% for a final read through to **check** and tidy-up errors.

How will you be assessed?

AS Level

There are two exams for AS Level Geography. Unlike A Level there is no coursework – but you must complete two days of AS fieldwork. The AS exams will involve:

Paper 1: Physical geography and people and the environment (1 hour 30 minutes)	Paper 2: Human geography and geography fieldwork investigation (1 hour 30 minutes)
There are a total of 80 marks available, worth 50% of the AS qualification. The paper consists of multiple-choice, short-answer levels of response and includes 9- and 20-mark extended prose questions. • Section A: **one** of *Water and carbon cycles* **or** *Coastal systems and landscapes* **or** *Glacial systems and landscapes* – **40 marks** • Section B: **either** *Hazards* **or** *Contemporary urban environments* – **40 marks**	There are a total of 80 marks available, worth 50% of the AS qualification. The paper consists of multiple-choice, short-answer levels of response and includes 9- and 20-mark extended prose questions. • Section A: *Changing places* – **40 marks** • Section B: *Fieldwork skills* – **40 marks**

A Level

There are two exams for A Level Geography and one individual fieldwork investigation – you must complete at least four days of A Level fieldwork. The A Level assessment will involve:

Paper 1: Physical geography 120 marks (40%)	Paper 2: Human geography 120 marks (40%)	Paper 3: Geography fieldwork investigation 60 marks (20%)
The 2 hours 30 minutes paper consists of multiple-choice, short-answer levels of response and includes 20-mark extended prose questions. • Section A: *Water and carbon cycles* – **36 marks** • Section B: **one** of *Hot desert systems and landscapes* **or** *Coastal systems and landscapes* **or** *Glacial systems and landscapes* – **36 marks** • Section C: **either** *Hazards* **or** *Ecosystems under stress* – **48 marks**	The 2 hours 30 minutes paper consists of multiple-choice, short-answer levels of response and includes 20-mark extended prose questions. • Section A: *Global systems and global governance* – **36 marks** • Section B: *Changing places* – **36 marks** • Section C: **either** *Contemporary urban environments* **or** *Population and the environment* **or** *Resource security* – **48 marks**	You must complete an individual investigation based on a question or issue defined and developed by you relating to any part of the specification content. You are advised to write between 3000 and 4000 words. Your investigation will be marked by your teachers and moderated by AQA.

How will your exam papers be marked?

Examiners have to know what it is that they are assessing you on, so they use Assessment Objectives, or AOs for short. There are three AOs for AS and A Level.

- **AO1: Demonstrate knowledge and understanding** of places, environments, concepts, processes, interactions and change, at a variety of scales (30–40%).
- **AO2: Apply knowledge and understanding** in different contexts to interpret, analyse and evaluate geographical information and issues (30–40%).
- **AO3: Use a variety of relevant quantitative, qualitative and fieldwork skills** to:
  - investigate geographical questions and issues
  - interpret, analyse and evaluate data and evidence
  - construct arguments and draw conclusions (20–30%).

Oxford's *AQA Geography A Level and AS Exam Practice and Skills* book will help you make sense of all this. It includes essential guidance on effective examination techniques, such as planning and structuring your longer answers and essays, e.g. **BUD**ing (**B**ox the command, **U**nderline the key words and **D**issect the question) and **PEEL**ing. Furthermore, at A Level it will help you to:

- understand the geographical themes/concepts that run across the course (Figure **9**)
- think about the course as a whole and how to approach it synoptically
- respond to questions about links between topics (or between sub-topics) more effectively.

Understanding the mark schemes

There are two types of mark scheme: point marked (up to 4 marks); and level marked (for 6 marks or more).

Point marking – short answers (worth up to 4 marks)

Questions carrying up to 4 marks are **point marked**. For every correct point that you make, you earn a mark. Sometimes these are single marks for a 1-mark question. Others require the development of points for additional marks. For example, if you are asked to describe one feature of something for 3 marks, development will be required.

> Outline **the role of wind in** affecting coastal energy. **[3 marks]**

There are three marks for this question and you have to outline the role of wind. You would receive one mark per valid point, for example:

'wind is responsible for the generation of waves (√) as friction occurs at the surface of the water' (√)

with additional credit for development, for example:

this has a direct bearing upon the potential for longshore drift depending upon the angle that the waves hit the coastline' (√).

Level marking – extended answers (worth 6 marks or more)

Questions carrying 6 marks or more are **level marked**. The examiner reads your whole answer and then uses a set of criteria – known as levels – to judge its qualities. There are two levels for questions carrying 6 marks, three levels for 9 marks, and four levels for 20 marks.

Think of levels in a mark scheme as a staircase you have climb in order to reach the top. Each step up to the next level, demands a little bit more of you (Figure **8**).

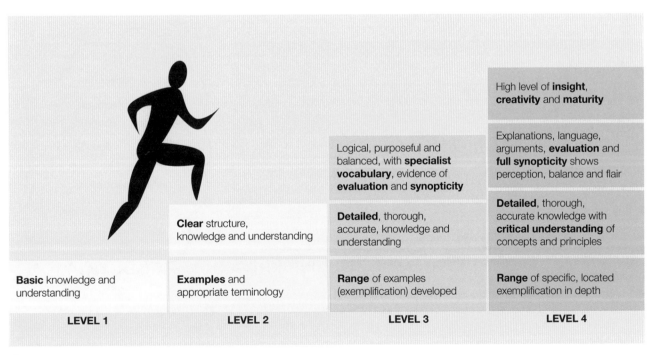

High level of **insight, creativity** and **maturity**

Explanations, language, arguments, **evaluation** and **full synopticity** shows perception, balance and flair

Logical, purposeful and balanced, with **specialist vocabulary**, evidence of **evaluation** and **synopticity**

Detailed, thorough, accurate knowledge with **critical understanding** of concepts and principles

Clear structure, knowledge and understanding

Detailed, thorough, accurate, knowledge and understanding

Basic knowledge and understanding

Examples and appropriate terminology

Range of examples (exemplification) developed

Range of specific, located exemplification in depth

| LEVEL 1 | LEVEL 2 | LEVEL 3 | LEVEL 4 |

🔺 *Figure 8* *Example steps in level marking*

Figure **8** shows general principles of level marking, but the actual criteria – determined by the AOs – will vary according to the nature of the question set. Hence the importance of examining and discussing mark schemes with your teachers.

For example, 9-mark questions have three levels in their mark schemes. The balance of marks is split between 4 marks for AO1 and 5 marks for AO2. So to achieve Level 3 you should aim to:

- demonstrate accurate knowledge and understanding throughout (AO1)
- apply your knowledge and understanding (AO2)
- produce a full interpretation that is relevant and supported by evidence (AO2)
- make supported judgements in a balanced and coherent argument (AO2).

Mark schemes for 20-mark questions have four levels. The balance of marks is split evenly with 10 marks for AO1 and 10 marks for AO2. So to achieve Level 4, in addition to the qualities for Level 3 you should aim to:

- reach a detailed evaluative conclusion that is rational and firmly based on knowledge and understanding which is applied to the context of the question (AO2)
- show a detailed, coherent and relevant analysis and evaluation in the application of knowledge and understanding throughout (AO2)
- show full evidence of links between knowledge and understanding and the application of knowledge and understanding in different contexts (AO2).

Examples and case studies

Case studies are in-depth examples of particular places, used to illustrate big ideas at localised scales. Unlike GCSE, few questions at AS and A Level specifically ask for examples, but you can only access higher levels of the mark schemes by including them to help you better demonstrate the points you make.

The important thing is that discussion of located, named examples makes your answer precise. But be selective – no question at this level will ever ask you to write everything you know about a case study.

Synopticity

Synopticity refers to your capability to demonstrate a comprehensive understanding of the whole picture. In effect it is demonstrated when your exam answers show evidence of the complete geographer – the capacity to draw connections and supporting evidence from *anywhere* in the course. Hence its inclusion in the highest levels of level marking.

Indeed, the higher tariff questions are likely to ask you to make **links**. For example:

- **links** between subtopics within a major topic, such as Water and carbon cycles
- **links** between elements of two different topics such as *Physical causes of changes in the carbon cycle* (Water and carbon cycles) and *Fires in nature* (Hazards)
- **links** between topics that you have learnt and novel situations or phenomena that are not in the specification; these type of questions will come with a resource and you'll have to apply what you know to answer the question.

Finally, don't fret about synopticity. With thorough revision and practice you will get there!

thresholds
causality equilibrium
systems feedback
representation
inequality identity
globalisation
interdependence
mitigation adaptation
sustainability
resilience risk

⊘ *Figure 9* The key geographical themes at AS and A Level. Can you define each of these themes or concepts? Can you give real world examples for more than one topic?

Exam question command words

These quite simply tell you what to do. You cannot answer questions properly if you don't understand them – so when you first read a question, check out the command word. It is *crucial* to ensure that you recognise and understand these commands instinctively before you sit your examinations.

Command word	What is it asking you to do?	Assessment objective
Analyse	Break down concepts, information and/or issues to convey an understanding of them by finding connections and causes and/or effects.	AO2 and AO3
Annotate	Add to a diagram, image or graphic a number of words that describe and/or explain features, rather than just identify them (which is labelling).	AO3
Assess	Consider several options or arguments and weigh them up so as to come to a conclusion about their effectiveness or validity.	AO1 – but mainly AO2
Compare	Describe the similarities and differences of at least two phenomena.	AO1 or AO3
Contrast	Point out the differences between at least two phenomena.	AO1 or AO3
Critically	Often occurs before 'Assess' or 'Evaluate' inviting an examination of an issue from the point of view of a critic with a particular focus on the strengths and weaknesses of the points of view being expressed.	AO1 – but mainly AO2
Define..., What is meant by...	State the precise meaning of an idea or concept.	AO1
Describe	Give an account in words of a phenomenon which may be an entity, an event, a feature, a pattern, a distribution or a process. For example, if describing a landform say what it looks like, give some indication of size or scale, what it is made of, and where it is in relation to something else (field relationship).	AO1
Distinguish between	Give the meaning of two (or more) phenomena and make it clear how they are different from each other.	AO3
Evaluate	Consider several options, ideas or arguments and form a view based on evidence about their importance/validity/merit/utility.	AO1 – but mainly AO2
Examine	Consider carefully and provide a detailed account of the indicated topic.	AO1
Explain... Why... Suggest reasons for...	Set out the causes of a phenomenon and/or the factors which influence its form/nature. This usually requires an understanding of processes.	AO1 and AO2
Interpret	Ascribe meaning to geographical information and issues.	AO3
Justify	Give reasons for the validity of a view or idea or why some action should be undertaken. This might reasonably involve discussing and discounting alternative views or actions.	AO1 – but mainly AO2
Outline..., Summarise...	Provide a brief account of relevant information.	AO1
To what extent...	Form and express a view as to the merit or validity of a view or statement after examining the evidence available and/or different sides of an argument.	AO1 – but mainly AO2

1 Water and carbon cycles

Your exam

AL *Water and carbon cycles* is a **core topic**. You must answer **all** questions in Section A: Water and carbon cycles as part of Paper 1: Physical geography.
Paper 1 carries 120 marks and makes up 40% of your A Level. Section A carries 36 marks.

AS *Water and carbon cycles* is an **optional topic**. You must answer **one** question in Section A of Paper 1: Physical geography and people and the environment, from a choice of three: *Water and carbon cycles* **or** *Coastal systems and landscapes* **or** *Glacial systems and landscapes*.
Paper 1 makes up 50% of your AS Level. Section A carries 40 marks.

Specification subject content
(Specification reference in brackets)

Either tick these boxes as a record of your revision,
or use them to identify your strengths and weaknesses

Your revision checklist

Section in student book and revision guide	☹	☺	☺	Key terms you need to understand Complete the **key terms** (not just the words in bold) as your revision progresses. 1.1 has been started for you.
Water and carbon cycles as natural systems (3.1.1.1)				
1.1 Systems in physical geography				*open system, inputs, closed system*
The water cycle (3.1.1.2)				
1.2 The global water cycle				
1.3 Changes in the magnitude of the water cycle stores				
1.4 The drainage basin system				
1.5 The water balance				
1.6 The flood hydrograph				
1.7 Factors affecting changes in the water cycle				

The carbon cycle *(3.1.1.3)*				
1.8 The global carbon cycle: stores				
1.9 The global carbon cycle: transfers				
1.10 Changes in the carbon cycle: physical causes				
1.11 Changes in the carbon cycle: human causes				
1.12 The carbon budget				
Water, carbon, climate and life on Earth *(3.1.1.4)*				
1.13 Water, carbon and climate change				
1.14 Mitigating the impacts of climate change				
1.15 Tropical rainforests: the water cycle				
1.16 Tropical rainforests: the carbon cycle				
Case studies *(3.1.1.6)*				
1.17 River catchment: the River Exe, Devon				
1.18 River catchment field data: the River Exe, Devon				

Note: The 'missing' specification reference **3.1.1.5** refers to skills

You need to know:
- what a system is
- the terminology associated with systems.

Student Book
pages 8–9

What is a system?

An ecosystem (ecological **system**) describes the interrelationships between living and non-living components within a particular environment, such as a pond (Figure 1). A diagram can be used to show parts (components) of a system and the **flows/transfers** between them.

A systems approach helps us to understand the physical and human world around us – they may be **open** (linked to other systems) or **closed** (entirely self-contained). We can apply this approach to physical systems (such as drainage basins) or to human systems (such as the operations in a factory).

Big idea

A system comprises stores/components with flows or transfers between them.

Inputs include:
- precipitation
- leaf fall during the autumn
- seeds carried by wind and birds

Stores/components include:
- water
- soil
- plants

Outputs include:
- water soaking through soil and rocks
- evaporation
- seed dispersal

Flows/transfers include:
- photosynthesis
- infiltration
- transpiration

▲ **Figure 1** *The components of a garden pond ecosystem*

Systems term	Definition	Drainage basin example
Input	Material or energy moving into the system from outside	Precipitation
Output	Material or energy moving from the system to the outside	Runoff
Energy	Power or driving force	Latent heat associated with changes in the state of water
Stores/ components	The individual elements or parts of a system	Trees, puddles, lakes, soil
Flows/transfers	The links or relationships between the components	Infiltration, groundwater flow, evaporation
Positive feedback	A cyclical sequence of events that amplifies or increases change	Rising sea levels can destabilise ice shelves → increasing the rate of calving → leading to an increase in melting → causing sea levels to rise further
Negative feedback	A cyclical sequence of events that damps down or neutralises the effects of a system, promoting stability and a state of dynamic equilibrium	Increased surface temperatures increase evaporation → which condenses at altitude to increase cloud cover → which reflects radiation → which cools the surface
Dynamic equilibrium	This represents a state of balance within a constantly changing system	Remote and unaffected drainage basin where there has been no significant natural or human impacts, or one that has had time to adjust to change

▲ **Figure 2** *Systems terminology*

Sixty second summary

- A system illustrates the interrelationships between components in the natural world.
- Key terms such as inputs, outputs, stores and transfers are used to describe different aspects of a natural system.
- Systems can be described as being 'open' (e.g. drainage basins) or 'closed' (e.g. the global water cycle).

Over to you

Write your own definitions of inputs, outputs, stores and transfers, giving examples for **each**.

Student Book
pages 10–11

You need to know:
- the principle operations of the water cycle
- the water cycle stores and their global distribution.

What is the water cycle?

Figure **1** shows the global water cycle – water is transferred between stores in a closed system.

- Stores – most of the Earth's water is saltwater, in the oceans. Ice sheets and groundwater are the main freshwater stores.
- Transfers – the processes involved in transferring water between stores (e.g. precipitation transfers water from the atmosphere to the Earth's surface).

Main stores in the water cycle

Water is stored within four major physical systems – the **lithosphere** (land), **hydrosphere** (liquid water), **cryosphere** (frozen water) and **atmosphere** (air).

- 97.4% of Earth's water is saline (Figure **2**).
- Only 2.5% is freshwater, stored mainly as snow and ice and groundwater.
- Surface and other freshwater comprises only 1.2% of all freshwater.

Global distribution of water stores

The distribution of land, sea, ice sheets, rivers, lakes and groundwater aquifers has a profound impact on global distribution of water.

Groundwater aquifers

Just over 30% of all freshwater is stored in *aquifers* – water stored in porous, permeable rocks (such as chalk and sandstone). The upper level of saturated rock is the *water table*, which rises and falls in response to groundwater flow, water abstraction or by *recharge* (additional water flowing into the rock).

Ancient (*fossil*) aquifers in Africa, the Middle East and Australia were formed thousands of years ago when the climate was much wetter.

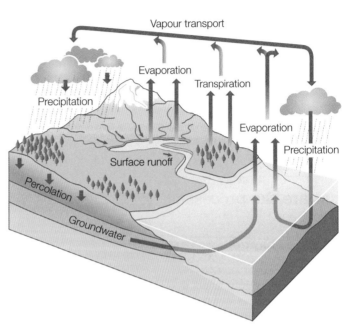

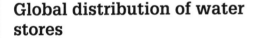

🔺 **Figure 1** *Stores and transfers of water in the global water cycle; no water is lost or gained globally*

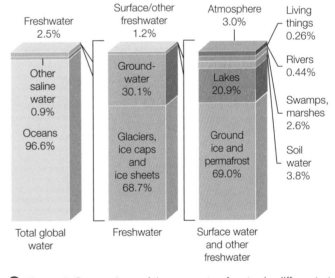

🔺 **Figure 2** *Comparison of the amounts of water in different stores*

⏱ **Sixty second summary**

- The water cycle essentially describes the Earth's stores and transfers of water.
- The vast majority of water is stored in the oceans as saltwater.
- Most freshwater is stored as ice or as groundwater within rocks (aquifers).
- Water is stored unevenly across the world due to the uneven distribution of both land/sea and also of permeable/porous rock which enable aquifers to form.

✏ **Over to you**

Using column 3 of Figure **2**, create a pie chart showing the composition of surface water and other freshwater on the Earth.

You need to know:
- how long water is stored in the components of the water cycle
- transfer processes which govern changes in magnitude
- processes of change at the global and local scale.

*Student Book
pages 12–15*

How long does water remain in the water cycle stores?

Figure **1** shows the typical residence times that water remains in each store. These varying time scales are extremely important in understanding the complexity of transfers within the water cycle.

Figure 1 ◗
Typical residence times of water found in various stores

Groundwater (shallow)	100–200 years
Groundwater (deep)	10 000 years
Glaciers	20–100 years
Seasonal snow cover	2–6 months
Lakes	50–100 years
Soil water	1–2 months
Rivers	2–6 months

What are the processes of change?

1 Climate change

At the peak of the last Ice Age (about 18 000 years ago), about a third of the Earth's land area was covered by glaciers and ice sheets, increasing the magnitude of the snow and ice store significantly. With less liquid water reaching the oceans, sea levels were over 100 m lower than the present day.

2 Cloud formation and the causes of precipitation

The driving force behind cloud formation and precipitation is the global atmospheric circulation model (Figure **3**).

- At the Equator, high temperatures result in high rates of evaporation.
- The warm, moist air rises, cools and condenses to form towering banks of cloud and heavy rainfall in a low pressure zone called the Inter-Tropical Convergence Zone (ITCZ).
- Seasonally, this zone moves north and south, with the overhead Sun illustrating both the spatial and temporal changes in transfers and store magnitudes that occur within the water cycle.

3 Cryospheric processes

Frozen water (ice) is the second largest store of water – 95% of which is locked up in the ice sheets covering Antarctica and Greenland.

The melting of all the polar ice sheets could result in a 60 m rise in sea level, adding a great deal of water to the ocean store. Rising sea levels destabilise ice shelves, triggering calving and further melting. This is an example of a positive feedback.

Process (flows/transfers)	Definition
Precipitation	Transfer of water from atmosphere to ground, in the form of rain, snow, hail, dew.
Evaporation/ Evapotranspiration	Transfer of water from liquid to gaseous state (water vapour), mostly from oceans to atmosphere.
Condensation	Transfer of water from gaseous to liquid state, e.g. formation of clouds.
Sublimation	Transfer from solid (ice) to gaseous state (water vapour) and vice versa.
Interception	Water intercepted and stored on leaves of plants.
Overland flow	Transfer of water over the land surface.
Infiltration	Transfer of water from the ground surface into soil where it may then percolate into underlying rocks
Throughflow	Water flowing through soil towards a river channel.
Percolation	Water soaking into rocks.
Groundwater flow	Very slow transfer of water through rocks.

▲ *Figure 2* Processes responsible for the changes in water stores

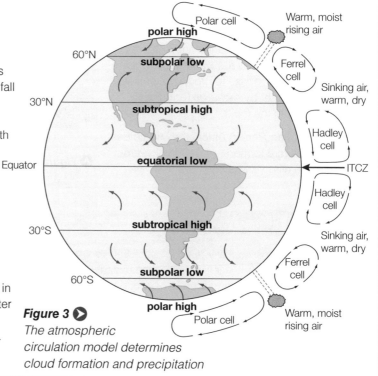

Figure 3 ◗
The atmospheric circulation model determines cloud formation and precipitation

Processes of change at the local scale

The hillslope water cycle

The magnitude of the stores in a hillslope water cycle changes in response to a wide variety of processes, such as *infiltration* – the movement (transfer) of water from the ground surface into the soil.

Water that is able to infiltrate the soil may be stored for very long periods of time either in the soil or deep within the underlying bedrock.

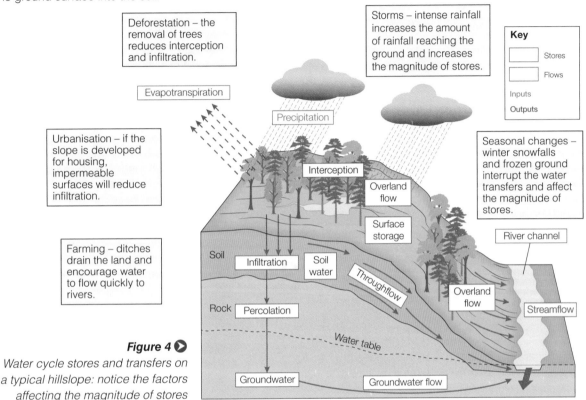

Deforestation – the removal of trees reduces interception and infiltration.

Storms – intense rainfall increases the amount of rainfall reaching the ground and increases the magnitude of stores.

Urbanisation – if the slope is developed for housing, impermeable surfaces will reduce infiltration.

Seasonal changes – winter snowfalls and frozen ground interrupt the water transfers and affect the magnitude of stores.

Farming – ditches drain the land and encourage water to flow quickly to rivers.

Figure 4 ▶
Water cycle stores and transfers on a typical hillslope: notice the factors affecting the magnitude of stores

The **soil water budget** describes the changes in the soil water store during the course of a year (Figure **5**). It depends on the type and depth of the soil, its texture and permeability.

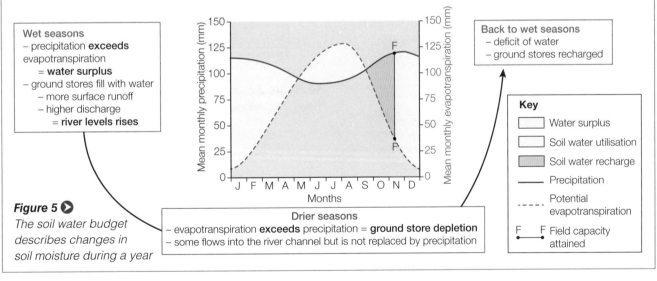

Wet seasons
– precipitation **exceeds** evapotranspiration
  = **water surplus**
– ground stores fill with water
  – more surface runoff
  – higher discharge
  = **river levels rises**

Back to wet seasons
– deficit of water
– ground stores recharged

Drier seasons
– evapotranspiration **exceeds** precipitation = **ground store depletion**
– some flows into the river channel but is not replaced by precipitation

Figure 5 ▶
The soil water budget describes changes in soil moisture during a year

Sixty second summary

- Water is stored for varying amounts of time in water cycle stores.
- Changes in magnitude involves transfer processes, such as evaporation.
- Several factors, such as climate change, drive change and variability in store magnitudes at a global scale.
- At the local scale, processes of change can be identified within the hillslope water cycle and the soil water budget.

Over to you

Use a series of flashcards to summarise the key processes of change and variability in the magnitude of water stores.

You need to know:

- terminology associated with the drainage basin
- key features of the drainage basin system (inputs, outputs, stores and transfers).

Student Book
pages 16–17

What is the drainage basin system?

A drainage basin is the area of land that is drained by a river and its tributaries (Figure **1**).

The movement of water within the drainage basin is illustrated by the drainage basin hydrological cycle or the *drainage basin system* (Figure **2**). This is an open system, with inputs (precipitation) and outputs (runoff, evapotranspiration).

Precipitation may be intercepted and stored by plants and trees before being evaporated. On the surface, it is stored as puddles, flows over the ground (overland flow) or infiltrates the soil.

Groundwater flow is a slow method of transfer, which feeds rivers through their banks and bed. It continues to supply water well after an individual rainfall event has occurred.

Infiltration capacity (rate of infiltration) is exceeded when the soil is unable to absorb water at the rate at which it is falling (or melting). Thin, frozen or already saturated soils will usually have a low infiltration capacity.

Overland flow is when water is unable to infiltrate and runs off the surface, flowing across a large surface area (*sheetflow*) or concentrated into small channels called *rills*.

Throughflow is when water passes through soils (rather than being stored as **soil water**). Coarse, sandy soils have a *low field capacity* (retain little water with rapid transfer). Clay soils drain more slowly and have a *high field capacity*.

Water passes through the soil until it reaches the water table (the upper level of saturated ground) or the underlying bedrock.

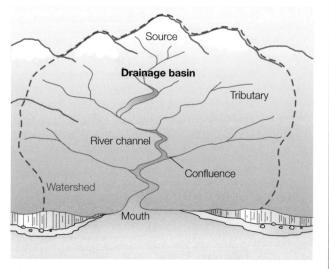

Figure 1 *Terminology associated with a drainage basin*

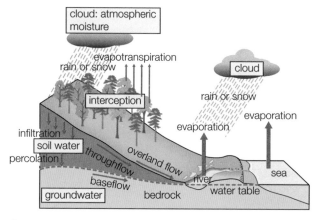

Figure 2 *Stores and flows within the drainage basin hydrological cycle*

Read about the role of vegetation on page 17 of the student book.

 Sixty second summary

- The drainage basin is an 'open' system with inputs (e.g. precipitation) and outputs (e.g. runoff)
- There are many transfer processes (e.g. infiltration, throughflow), which affect the magnitude of stores (e.g. soil, rock, vegetation).

 Over to you

Make a simple copy of the drainage basin hydrological cycle. Add annotations to outline some of the key characteristics.

Student Book
pages 18–19

You need to know:

- the water balance equation
- factors affecting variations in runoff, for example, rock type, land use and soil moisture.

What is the water balance?

The **water balance** helps understanding of the characteristics of individual drainage basins.
The water balance is expressed as:

$$P = O + E +/- S$$

where P = precipitation,   O = total runoff (streamflow),

E = evapotranspiration,   S = storage (in soil and rock)

Read about the calculation of the water balance of the River Wye on pages 18–19 of the student book.

What causes variations in runoff?

The total runoff (expressed as a percentage of precipitation) is a measure of the proportion of the total precipitation that makes its way into streams and rivers (Figure **1**).

The type and intensity of precipitation are important.

- Intense rainfall is more likely to pass quickly into rivers, increasing runoff.
- Drizzle will be held in the trees and on the grass, increasing evaporation.
- Snow will delay any runoff, but when frozen soils melt, runoff values might be high.

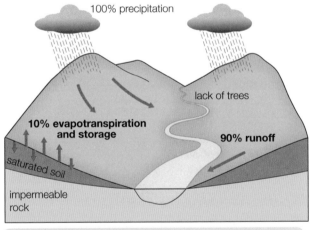

100% precipitation

lack of trees

10% evapotranspiration and storage

90% runoff

saturated soil

impermeable rock

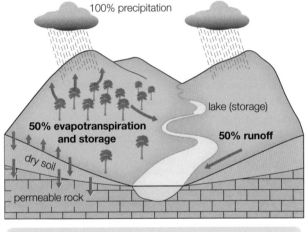

100% precipitation

lake (storage)

50% evapotranspiration and storage

50% runoff

dry soil

permeable rock

Runoff is high (90%), so most precipitation is transferred straight to the river – little is lost or stored on the way. Flooding is likely.

Runoff is lower (50%). Dense forests and permeable rocks are reasons why a high proportion of precipitation is lost or stored before it reaches the river. Flooding is less likely.

 Figure 1 Variations in land use, geology and infiltration capacity can affect runoff

 Sixty second summary

- The water balance equation is P = O + E +/– S.
- Runoff, expressed as a percentage of precipitation, enable comparisons to be made between drainage basins.
- Several factors affect runoff – rock type (permeable/impermeable), soil moisture and land use.
- High percentages of runoff are associated with saturated soils, impermeable rocks and lack of plant cover (e.g. urbanisation, deforestation).

Over to you

Outline the main factors affecting the water balance of a drainage basin and illustrate (exemplify) **each** one.

The flood hydrograph

The **flood** (storm) **hydrograph** is a graph showing the discharge of a river following a particular storm event (Figure **1**). Despite the unique nature of river hydrographs, it is possible to identify two models representing polar opposites (Figure **2**).

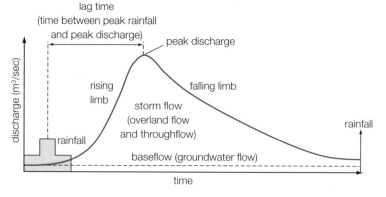

Figure 1 ▷
Terminology of a flood hydrograph

What is discharge?

Discharge is the volume of the water flowing down a river. Values for river discharge are expressed in 'cumecs' (cubic metres per second). It can be expressed as:

Discharge (m³ per second) = cross-sectional area (m²) × velocity (metres per second)

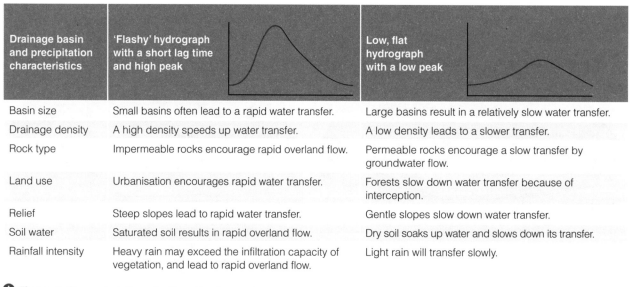

Drainage basin and precipitation characteristics	'Flashy' hydrograph with a short lag time and high peak	Low, flat hydrograph with a low peak
Basin size	Small basins often lead to a rapid water transfer.	Large basins result in a relatively slow water transfer.
Drainage density	A high density speeds up water transfer.	A low density leads to a slower transfer.
Rock type	Impermeable rocks encourage rapid overland flow.	Permeable rocks encourage a slow transfer by groundwater flow.
Land use	Urbanisation encourages rapid water transfer.	Forests slow down water transfer because of interception.
Relief	Steep slopes lead to rapid water transfer.	Gentle slopes slow down water transfer.
Soil water	Saturated soil results in rapid overland flow.	Dry soil soaks up water and slows down its transfer.
Rainfall intensity	Heavy rain may exceed the infiltration capacity of vegetation, and lead to rapid overland flow.	Light rain will transfer slowly.

▲ **Figure 2** *Characteristics of a flood hydrograph*

 Sixty second summary

- Discharge is a measurement of the river's volume over time, expressed in cubic metres per second.
- A flood hydrograph shows the response of a river to a storm event through variations in its shape.
- The shape of a flood hydrograph reflects drainage basin characteristics (such as steepness of land, land use and soil moisture) as well as precipitation characteristics.

 Over to you

Draw a simple hydrograph for a 'flashy' drainage basin. Label the following components: peak discharge, lag time, rising limb, falling limb. Outline the likely drainage basin characteristics likely to lead to this shape.

Student Book
pages 22–3

You need to know:
- how natural change impacts on the water cycle
- how human activities impact on the water cycle.

Natural (physical) variations affecting change

Extreme weather events can have significant impacts on the water cycle. For example, in times of drought:

- Water stores in rivers are reduced and throughflow ceases
- Water stores in rivers and lakes are reduced
- The soil water store is reduced as soils dry out
- Vegetation dies off affecting transpiration, interception and infiltration
- Groundwater flow becomes more important – it is a long-term transfer and not affected by short-term weather extremes
- Dry air causes initial high rates of evapotranspiration but reduces as water on the ground dries up

▲ **Figure 1** *Castiac Lake, California at half its usual capacity, 2014*

Seasonal changes can also impact the water cycle (Figure **2**).

Water cycle component	Summer	Winter
Precipitation	Total rainfall may be less but storms are more frequent.	Greater quantities of rainfall with a likelihood of snow.
Vegetation – interception, transpiration, etc	Vegetation grows rapidly increasing interception and transpiration.	Vegetation dies back reducing interception and transpiration.
Evaporation	Higher temperatures encourage rapid evaporation (warm air can hold more moisture).	Lower temperatures reduce rates of evaporation.
Soil water	Dry soils encourage infiltration. But hard, baked soils encourage overland flow.	Soils may become saturated, leading to overland flow.
River channel flow	Low flow conditions are more likely.	High flow conditions are more likely.

▲ **Figure 2** *This table illustrates seasonal impacts on the UK water cycle*

Human activities affecting change

Land-use change

- Water cannot infiltrate impermeable concrete and tarmac, so *urbanisation* increases overland flow and soil water and groundwater stores are reduced. Flooding is more likely.
- *Deforestation* leads to surface runoff and soil erosion and reduced soil water stores.

Farming practices

Farmers can control the local water cycle through irrigation or drainage. Soils covered with plants have higher infiltration rates, and, therefore, reduced runoff. **Desertification** reduces the capacity of soil to retain water.

Water abstraction

Excessive abstraction can have significant effects on the local water cycle. Aquifers can become depleted or contaminated by inflowing saltwater if the water table drops below sea level.

Peatland drainage in the UK

The drainage of peatlands in the East Anglian Fens transformed submerged landscape to highly productive farmland. However, this can have significant impacts on both the water cycle and the carbon cycle – the water table is lowered, changing rates of infiltration and evaporation.

Dry peat is friable and vulnerable to erosion. Peatlands are composed of partly decomposed vegetation and are important carbon stores. Vegetation on top of the peat also absorbs carbon dioxide (CO_2) from the atmosphere. As the peatlands are drained, air penetrates deeper, enabling decomposition of the carbon, releasing CO_2.

 Sixty second summary

- Natural factors can affect the water cycle (e.g. drought, seasonal variations in evapotranspiration and soil water).
- Human activities can affect the water cycle (e.g. land-use change, farming practices and water abstraction).

 Over to you

Summarise how seasonal variations affect precipitation, evaporation and soil water in the UK water cycle.

Student Book
pages 24–5

You need to know:

- the global carbon cycle systems diagram
- key stores and transfer processes within the carbon cycle
- the global pattern of vegetation carbon storage.

The global carbon cycle

Carbon is a basic chemical element that, along with nitrogen, phosphorous and sulphur, is needed by all plants and animals in order to survive. The recycling of carbon is essential for life on Earth (Figure **1**).

- *Stores* – the main stores of carbon are the lithosphere (rocks and soil), hydrosphere (oceans), cryosphere (snow and ice), atmosphere and biosphere (plants). A **carbon sink** is a store that absorbs more carbon than it releases. A **carbon source** releases more carbon than it absorbs.
- *Transfers* – the processes involved in transferring carbon between the stores (e.g. photosynthesis takes carbon dioxide (CO_2) out of the atmosphere and converts it into carbohydrates, such as glucose within plants).

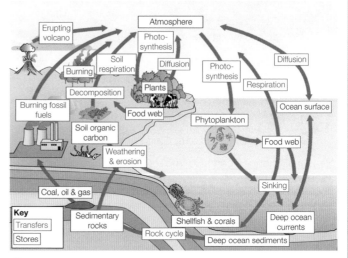

Figure 1 *The global carbon cycle showing the interrelationships between stores and transfers*

What are the main stores in the carbon cycle?

Carbon store	Amount in billions of tonnes	Description
Marine sediments and sedimentary rocks	100 000	The largest long-term store of carbon, with rocks taking millions of years to form.
Ocean	38 000	CO_2 is absorbed directly from the air and rivers discharge carbon carried in solution.
Fossil fuel deposits	4000	Hydrocarbons (e.g. coal, oil and gas) are important long-term stores of carbon. Since the industrial revolution, combustion has increased levels of atmospheric CO_2.
Soil organic matter	1500	Carbon from rotting vegetation can remain in soils for hundreds of years. Deforestation, land-use change and soil erosion can rapidly release the stored carbon.
Atmosphere	750	Atmospheric CO_2 has increased due to emissions from power stations, vehicles and deforestation.
Terrestrial plants	560	Plants convert energy from the Sun into carbohydrates that support life. Plants can store carbon for many years and transfer it to the soil.

Figure 2 *Carbon stores in the global carbon cycle*

The global pattern of vegetation carbon storage

Some regions (e.g. the Sahara Desert), have virtually no plant storage; others have flourishing vegetation growth (e.g. warm temperate environments). Carbon uptake is increasing in middle/ high latitudes of the northern hemisphere. Less carbon is being absorbed in the tropics and southern hemisphere, mainly due to drought.

Sixty second summary

- Carbon recycling is vital for all life on Earth.
- The global carbon cycle comprises stores (e.g. forests) and transfers (e.g. photosynthesis).
- The largest stores are marine sediments, rocks and oceans.
- A carbon sink is a store with a net gain of carbon; a carbon source is store with a net loss.
- Vegetation storage varies globally depending on climate and the distribution of land and sea.

Over to you

When revising don't underestimate the power of **PQ2R** – **P**review skimming and anticipation – **Q**uestion to identify the main theme – **R**ead the page carefully (your mind will look for answers) – **R**eview to check understanding and test recall.

Student Book
pages 26–9

You need to know:

- about carbon transfers at the local scale
- about the carbon cycle associated with the lithosere
- transfer processes operating in the carbon cycle.

The carbon cycle at the local scale

The tree in Figure **1** acts as a carbon store (wood is about 50% carbon).

A terrestrial carbon cycle

- When rock is exposed for the first time, it is vulnerable to processes of weathering.
- As the rock is slowly broken down, carbon may be released, often dissolved in water.
- Over time, vegetation (such as lichen and moss) grows on the bare rock and carbon exchange starts to take place, involving photosynthesis and respiration.
- As organic matter is added, a soil develops that can support a wider range of plants.
- Over hundreds of years, plant species become more diverse, benefiting from the supply of carbon in the soil.

This sequence of changes is called a *vegetation succession*.

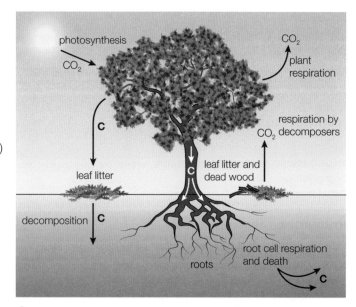

▲ **Figure 1** *Carbon cycle of a tree; several processes transfer carbon between the atmosphere and the land*

- A succession that relates to a specific environment is called a **sere**. Each stage in the succession is a **seral stage**. A **lithosere** is a vegetation succession that occurs on bare rock (as explained above). Other seres include **hydrosere** (fresh water), **halosere** (salt water), **psammosere** (sand dunes).
- The *climatic climax* is when a state of equilibrium is achieved.
- The climax vegetation for a lithosere in the UK will usually be a deciduous woodland (Figure **2**).

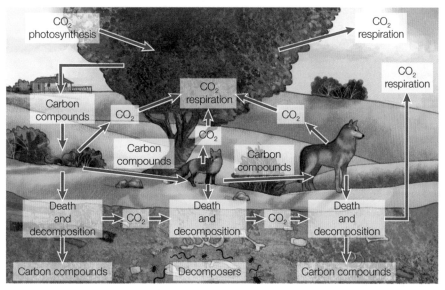

▲ **Figure 2** *A simple terrestrial (lithosere) carbon cycle; note the large number of interrelationships*

Continued over ➤➤➤

What are the main transfers operating in the carbon cycle?

Photosynthesis

Photosynthesis is the process whereby plants use light energy from the Sun to convert CO_2 from the air, and water from the soil into carbohydrates in the form of glucose (Figure **3**).

Respiration

Respiration is a chemical cellular process whereby glucose is converted into energy, which can be used for growth. CO_2 is then returned to the atmosphere, mostly by exhaled air.

Decomposition

When organisms die they are consumed by decomposers such as bacteria, fungi and earthworms. During **decomposition**, carbon from their bodies is returned as CO_2 to the atmosphere.

Combustion

Organic material contains carbon which, when burned in the presence of oxygen (e.g. coal in a power station), is converted into energy, CO_2 and water. This is **combustion**. The CO_2 is released into the atmosphere, returning carbon that might have been stored in rocks for millions of years.

Burial and compaction

Organic matter is buried by sediments and becomes compacted over time. Over millions of years, these organic sediments containing carbon may form *hydrocarbons* such as coal and oil.

Limestone formed from organic calcium-rich shells and coral, also stores large amounts of carbon.

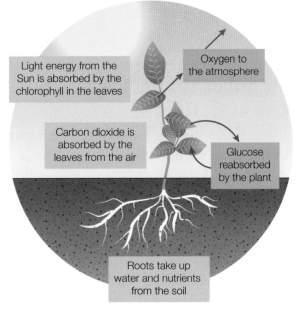

Figure 3 *The key features of photosynthesis*

Carbon sequestration

Carbon sequestration describes the transfer of carbon from the atmosphere to plants, soils, rock formations and oceans. *Carbon capture and storage (CCS)* is the technological 'capturing' of carbon emitted from power stations.

Weathering

When CO_2 is absorbed by rainwater it forms a mildly acidic carbonic acid. Through a series of complex chemical reactions, rocks will slowly dissolve with the carbon being held in solution. This process is called *carbonation*.

 Sixty second summary

- Carbon transfer processes can be identified at a very local scale (e.g. a tree).
- The lithosere is an example of a terrestrial carbon cycle, exhibiting stores and transfers as it develops.
- Several transfer processes operate in the carbon cycle (e.g. photosynthesis, respiration, decomposition).

 Over to you

From memory, write down the main transfer processes in the carbon cycle. Try to write a brief description of what **each** process involves. Then check your answers using the revision guide or student book.

You need to know:

- the effect of natural climate change on the carbon cycle
- the impacts of cold and warm periods on carbon stores and transfers
- the impacts of wildfires and volcanic activity on the carbon cycle.

Student Book
pages 30–3

Physical causes of changes in the carbon cycle

Impacts of natural climate change

During the Quaternary geological period (from 2.6 million years ago to the present day) global climates fluctuated considerably between warm (interglacial) and cold (glacial) periods (Figure **1**).

The causal relationship of the trends for CO_2 and temperature in Figure **1** is debatable. Increased CO_2 enhances the greenhouse effect, causing temperature to rise. However, increased temperature can lead to a rise in carbon, for example, due to melting permafrost releasing methane.

Impacts of cold (glacial) conditions

- Chemical weathering processes more active – cold water can hold more CO_2.
- Decomposers less effective, so reduced carbon transfer to soils.
- Less water reached the oceans, so less sediment transfer along rivers and less build-up of sediments on the ocean floor.
- Frozen soil over vast areas of land stopped transfers of carbon.

Impacts of warm (interglacial) conditions

Melting of permafrost in tundra regions, for example, in Siberia in Russia, has led to an increase in carbon emissions. This further enhances the greenhouse effect, leading to increased warming – an example of a positive feedback.

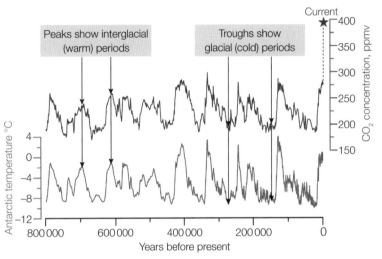

▲ *Figure 1 This graph is based on Antarctic ice core data. The peaks are warm, interglacial periods and the troughs are cold, glacial periods (ppmv=parts per million by volume).*

Impacts of wild fires

- Wild fires release CO_2 into the atmosphere, causing a spike in the rising trend of carbon emissions.
- Wild fires can turn forests from being a carbon sink to being a carbon source.

Impacts of volcanic activity

Volcanic activity releases CO_2 that has been trapped for millions of years in rocks beneath the crust (around 380 million tonnes per year. In comparison, human activities emit about 30 billion tonnes of CO_2 per year, mainly as a result of burning fossil fuels).

Sixty second summary

- Natural climate change affects stores and transfers, e.g. through variations in weathering, vegetation growth and water distribution.
- During the Quaternary period, CO_2 and temperature patterns followed the same trend, although any causal relationship is debatable.
- Cold periods are characterised by low carbon transfers due to frozen soils, limited decomposition and reduced flow of water.
- During warmer periods, melting permafrost increases carbon emissions (as CO_2 and methane).
- Wildfires and volcanic eruptions emit CO_2 into the atmosphere.

Read about the causal link between CO_2 and temperature on page 33 of the student book.

 Over to you

Outline the physical causes of change in the carbon cycle.

You need to know:

- the impacts of combustion of fossil fuels on the carbon cycle
- the effects of land-use change on the carbon cycle.

Student Book
pages 34–7

Human causes of changes in the carbon cycle

- The combustion of fossil fuels accounts for 90% of anthropogenic (human-related) carbon release.
- The remaining 10% results from land-use change, such as deforestation, land drainage and agriculture.
- Anthropogenic carbon is absorbed by oceans, vegetation and the atmosphere.
- Since the 1960s, global concentrations of CO_2 have increased dramatically from about 320 ppm to just over 400 ppm, the highest level ever recorded.

Combustion of fossil fuels

Today, most of the world's gas and oil is extracted from rocks that are 70–100 million years old. When burnt to generate energy and power, the stored carbon is released, primarily as CO_2 into the atmosphere (Figure **1**). Once in the atmosphere, it enhances the natural greenhouse effect, increasing global temperatures – so-called global warming.

Figure **2** shows the extent of global CO_2 increase in the atmosphere driven by industrialisation.

Land-use change

Land-use change is responsible for about 10% of carbon release globally. It impacts on relatively short-term stores and has direct links to issues of climate change and global warming.

Land-use changes, such as farming, deforestation and urbanisation can have a significant effect on small-scale carbon cycles.

Farming practices

Ploughing and harvesting, rearing livestock, using machinery fuelled by fossil fuels and using fertilisers based on fossil fuels are all farming practices that release carbon. On many farms it is the use of artificial fertilisers that is the main source of carbon emission.

Methane, a carbon compound, is a potent greenhouse gas. It is released from the cultivation of rice as well as belched from cattle.

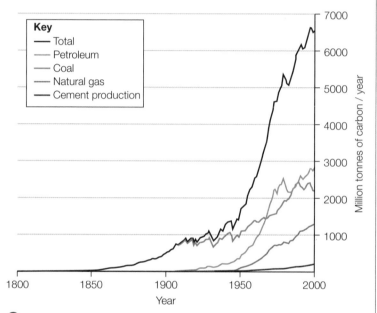

Figure 1 *The impact of industrialisation on carbon emissions*

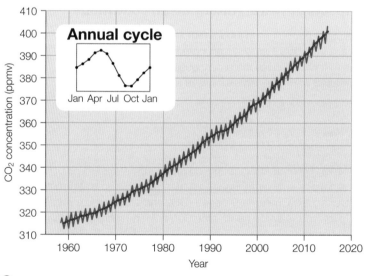

Figure 2 *Increase in CO_2 in the atmosphere since 1960*

See Figure **1** on page 34 of the student book for human impacts on the global carbon cycle.

Deforestation

Deforestation is concentrated in tropical regions, such as Indonesia. In total, it accounts for about 20% of all global CO_2 emissions.

When deforestation by burning occurs, carbon is immediately released into the atmosphere. If the land is then used for a different purpose, such as grassland for cattle ranching, the future absorption of CO_2 will be reduced, becoming a carbon source when deforested rather than a carbon sink (Figure **3**).

Urbanisation

Globally, urban areas occupy about 2% of the total land area. However, these areas account for 97% of all anthropogenic CO_2 emissions! The major sources of these emissions are transport, the development of industry, the conversion of land use from natural to urban, and cement production for the building sector.

Cement is used in construction across the world. Its production creates CO_2 as a by-product in the conversion of limestone to lime.

Look at the case study at Shimpling Park Farm on page 36 of the student book.

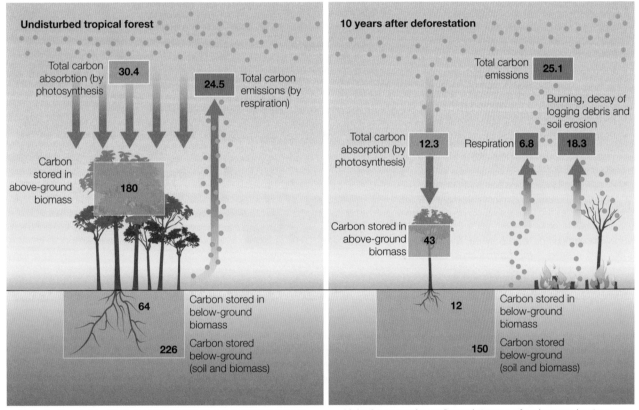

Units for fluxes (blue/red): tonnes of carbon per hectare per year

Units for stores (green/brown): tonnes of carbon per hectare

 Figure 3 *Changes in carbon cycling resulting from deforestation; notice that the area changes from being a carbon sink to a carbon source*

Sixty second summary

- The combustion of fossil fuels accounts for 90% of all human carbon emissions, enhancing the greenhouse effect.
- Farming practices, such as rearing livestock and the use of fertilisers and machinery, contribute to carbon emissions.
- Deforestation releases carbon stored in trees. Subsequent land uses may be less efficient carbon sinks in comparison.
- Urbanisation often replaces carbon sinks (grass/forest) and contributes to emissions through transport and industry (e.g. cement production).

Over to you

Create a spider diagram to summarise the key processes of change and variability in the magnitude of carbon stores.

Student Book
pages 38–9

You need to know:
- the definition of the carbon budget
- about major stores and transfers within the carbon budget
- the impacts of the carbon cycle on land, ocean and atmosphere.

What is the carbon budget?

The **carbon budget** uses data to describe the amount of carbon that is stored and transferred within the carbon cycle. It can be determined for all systems from a single tree to global (Figure **1**).

The impacts of the carbon cycle

- The carbon cycle releases CO_2 and other gases which absorb long-wave radiation warming the lower atmosphere – the *greenhouse effect*
- Vegetation removes CO_2 from the atmosphere and releases water and oxygen. Areas of dense vegetation therefore experience high rainfall.
- Plankton in the oceans may promote cloud formation, through the creation of dimethylsulphide (DMS).
- Ash and gases from volcanic eruptions absorb incoming radiation, cooling the Earth.

 Big idea

The **carbon budget** uses data to describe the amount of carbon that is stored and transferred within the carbon cycle.

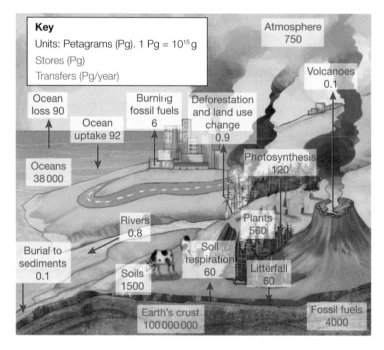

Key
Units: Petagrams (Pg). 1 Pg = 10^{15} g
Stores (Pg)
Transfers (Pg/year)

Atmosphere 750
Volcanoes 0.1
Ocean loss 90
Burning fossil fuels 6
Deforestation and land use change 0.9
Ocean uptake 92
Photosynthesis 120
Oceans 38 000
Rivers 0.8
Plants 560
Burial to sediments 0.1
Soil respiration 60
Litterfall 60
Soils 1500
Earth's crust 100 000 000
Fossil fuels 4000

Figure 1 ◗
The global carbon budget; most carbon is stored in the Earth's crust and oceans

Land	Ocean	Atmosphere
• The carbon cycle is responsible for the formation of soil. Carbon in the form of organic matter (litterfall) introduces important nutrients. • Organic matter is essential for plant growth and the production of food. • Carbon stored in grass provides fodder for animals. • Carbon provides a valuable source of energy, e.g. wood and fossil fuels.	• Carbon (in the form of calcium carbonate) is used by some marine organisms to build shells. • The carbon cycle has an impact on phytoplankton, a food for many marine organisms. • Phytoplankton consumes CO_2 during photosynthesis. The carbon is then passed along the marine food chain.	• CO_2 in the atmosphere helps to warm the Earth through the greenhouse effect. • Increases in carbon emissions as a result of human activities have led to the enhanced greenhouse effect, which threatens to have a profound impact on the world's climate. • Carbon stored by vegetation has a significant effect on the atmosphere, whether deforestation (carbon source) or afforestation (carbon sink).

▲ **Figure 2** *Some impacts of the carbon cycle on land, ocean and atmosphere*

 Sixty second summary

- The carbon budget describes the amount of carbon stored and transferred within the carbon cycle.
- Most carbon is stored in the Earth's crust. The oceans and fossil fuels are other major stores.
- Photosynthesis is the major process of carbon exchange.
- On land, carbon impacts on soil formation and on sources of energy.
- In oceans, calcium carbonate is used to build shells and corals.
- In the atmosphere, CO_2 forms an important greenhouse gas.

 Over to you

Make a list of the top **three** stores and transfers of carbon according to the carbon budget characteristics. Comment on the significance of these values.

Student Book
pages 40–3

You need to know:

- that water and carbon are essential in supporting life on Earth
- the relationship between the water cycle and the carbon cycle
- how feedback loops impact climate change.

The relationship between the water and carbon cycles

An important link between the water and carbon cycles is the ability of water to absorb and transfer CO_2 (Figure **1**):

1 Rainwater dissolves CO_2 to form weak carbonic acid.
2 Dissolved carbon is carried by rivers to the ocean.
3 Calcium carbonate in the ocean is used to form shells and coral.

Figure 1 ◗

Notice the importance of water in absorbing and transferring carbon

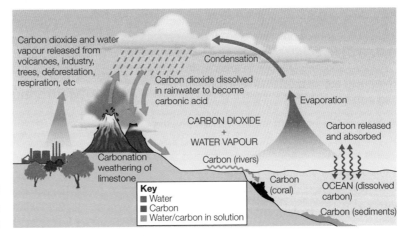

Links to climate change

Water cycle feedback loop

Absorption of the Sun's heat warms the oceans, which has reduced Arctic ice coverage alarmingly in recent years. Ice reflects radiation, so less ice cover means more heat is absorbed by the oceans – a positive feedback loop. These changes could affect water transfer between land, ocean and atmosphere and change patterns of precipitation.

Carbon cycle feedback loop

Warmer temperatures in the Arctic have two opposite effects:

- an extended plant-growing season increases carbon absorption from the atmosphere
- melting permafrost returns carbon to the atmosphere (Figure **2**).

Water cycle/carbon cycle feedback loop

Marine phytoplankton releases dimethylsulphide (DMS) that may promote the formation of clouds over the oceans. Warmer temperatures increase phytoplankton populations, which could lead to an increase in cloudiness and global cooling – a negative feedback loop.

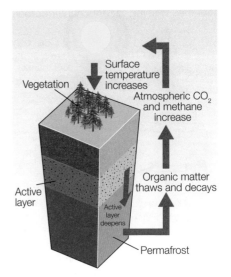

◗ **Figure 2** *Permafrost melting returns stored carbon back to the atmosphere – a positive feedback loop*

 Sixty second summary

- There are important links between the water cycle and the carbon cycle.
- The ability of liquid water to absorb and transfer carbon is crucial.
- Feedback loops involving melting permafrost/ice, vegetation and phytoplankton have significant impacts on climate change.

Over to you

Identify links (relationships) between the following:

- water cycle and carbon cycle
- water cycle and climate change
- carbon cycle and climate change
- water/carbon cycle and climate change.

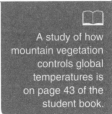

A study of how mountain vegetation controls global temperatures is on page 43 of the student book.

You need to know:

- how combustion, photosynthesis and deforestation can be modified by human actions
- the role of political initiatives in mitigating the impacts of climate change.

Student Book
pages 44–5

Which carbon transfers can be modified by human intervention?

Modifying industrial combustion

Carbon capture and storage (CCS) uses technology to capture CO_2 emissions from coal-fired power stations and industry. The gas is then transported to a site where it can be stored and prevented from entering the atmosphere. This could cut global carbon emissions by up to 19% (Figure **1**).

In 2014, Boundary Dam in Canada became the world's first commercial carbon capture coal-fired power plant. It aims to reduce greenhouse gas emissions by about 1 million tons a year, the equivalent of the emissions from 250000 cars.

Modifying photosynthesis

Plantation forests, which comprise an estimated 7% of the global forest area, are particularly effective in absorbing CO_2 compared to natural forests. For some time, this has been recognised by the IPCC as a legitimate option for countries wishing to reduce their carbon emissions.

Modifying land use

Apart from deforestation, farming practices are the most common cause of land-use change. *Carbon farming* is where one type of crop is replaced by another that can absorb more CO_2 from the atmosphere.

Modifying deforestation

Deforestation is a major cause of carbon emissions. There are several strategies aimed at reducing rates:

- Certification of timber products that have been grown sustainably by the Forestry Stewardship Council (FSC).
- Countries, organisations and individuals make carbon payments to offset their carbon emissions.
- In Malaysia, the Selective Management System is a sustainable approach to logging by felling selected trees and planting replacements.

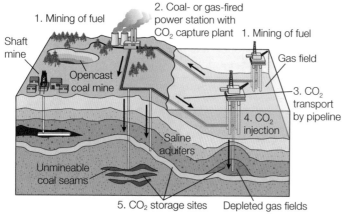

1. Mining of fuel
2. Coal- or gas-fired power station with CO_2 capture plant
Shaft mine
Opencast coal mine
1. Mining of fuel
Gas field
3. CO_2 transport by pipeline
4. CO_2 injection
Saline aquifers
Unmineable coal seams
5. CO_2 storage sites   Depleted gas fields

 Figure 1 *Carbon capture and storage; captured carbon is transported by pipeline to various stores deep underground. The stages of the process are numbered 1 to 5.*

Political initiatives: the Paris Agreement

At the climate conference in Paris (COP21) in 2015, 195 countries adopted the first universal legally binding global climate deal. It is due to be enforced by 2020. The agreement sets out an action plan to limit the average global temperature increase to 1.5 °C above pre-industrial levels.

Sixty second summary

- Human intervention in carbon transfers can mitigate the impacts of climate change.
- Carbon capture and storage can reduce the impacts of industrial combustion.
- Photosynthesis can be modified through the planting of forests.
- The rate of deforestation can be reduced by international agreements.
- Political initiatives can focus international action on reducing carbon emissions.

Over to you

Explain how modifications to combustion, photosynthesis and deforestation can mitigate the impact of climate change.

Student Book
pages 46–7

You need to know:

- the characteristics of the tropical rainforest water cycle
- the impacts of human activity and environmental change.

Characteristics of tropical rainforests

Tropical rainforests are found in a broad belt from Central and South America, through central parts of Africa, south-east Asia and into the northern part of Australia. They:

- have an annual rainfall of 2000+ mm with 27 °C temperatures throughout the year – ideal for plant growth
- are home to 200 million people and to about half of the world species of plants and animals
- absorb huge quantities of CO_2 and emit 28% of the world's oxygen.

What is the impact of human activity and environmental change?

Half of the world's rainforests have already been cleared for commercial farming, mining, logging and settlements. Deforestation wipes out some components and significantly affects others.

- Evapotranspiration is reduced, so the atmosphere less humid.
- Rainfall reaches the ground immediately, compacting it and encouraging overland flow.
- Exposed to the sun, the soil will become very dry and vulnerable to erosion.
- Very little interception of rainfall or evaporation, and transpiration will be virtually zero.
- Increased runoff, and risk of flooding.

Can deforestation affect climate and rainfall patterns?

Rainforests allow a considerable amount of water to be returned to the atmosphere through evapotranspiration. When forests are replaced by pasture or crops evapotranspiration is typically reduced, leading to reduced atmospheric humidity and suppressing precipitation.

The tropical rainforest water cycle

The dense forest canopy intercepts about 75% of rainfall. Some of this is evaporated and the rest drips or flows (**stemflow**) to the ground – to be used by plants, absorbed by the soil or lost from runoff.

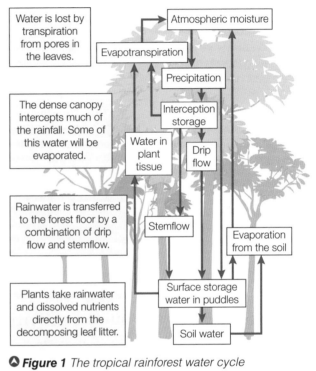

🔺 **Figure 1** The tropical rainforest water cycle

 Sixty second summary

- Tropical rainforests exhibit a distinctive water cycle, due to high precipitation, the extensive canopy and lush plant growth.
- Deforestation has profound impacts on the water cycle by exposing the soil to high impact rainfall.
- Deforestation may impact regional rainfall patterns by reducing moisture in the atmosphere.

Over to you

Outline the impacts of deforestation on the tropical rainforest water cycle. Make sure you include brief explanations (using accurate terminology) in your list.

You need to know:

- about the carbon cycle in tropical rainforests
- that human activity and environmental change impacts the carbon cycle.

Student Book
pages 48–51

The tropical rainforest carbon cycle

Figure **1** shows the carbon cycle operating in a tropical rainforest setting.

- The warm, wet climate is ideal for plant growth and photosynthesis, which absorbs huge quantities of CO_2.
- Wood is about 50% carbon, so rainforests are a huge carbon store and important 'carbon sinks' in mitigating the effects of global warming.
- Respiration by plants, trees and animals returns CO_2 to the atmosphere.
- Decomposition releases CO_2 back to the atmosphere.
- Some carbon may also be stored within the soil or dissolved and then removed by streams as an output from the rainforest system.

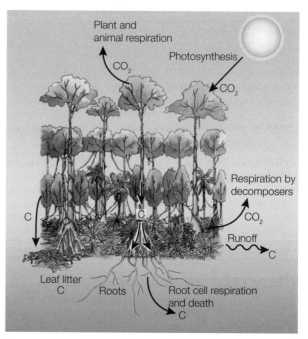

▲ **Figure 1** *The tropical rainforest carbon cycle*

What is the impact of human activity and environmental change?

Deforestation is one of the main contributors to the greenhouse gases that are responsible for climate change, principally through the release of CO_2 when the wood is burnt. Figure **2** shows other effects of deforestation.

Replacing rainforest with alternative land uses, such as crops and pasture, reintroduces stores and flows, although at much less effective levels.

Human activity in the rainforest can take place with minimal impact on the carbon cycle. For example, in Malaysia, there is strict regulation of logging followed by replanting, which is both sustainable and has little impact on the carbon cycle.

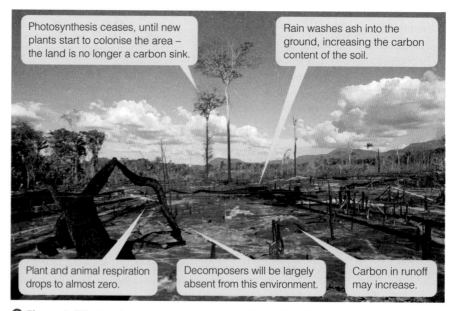

▲ **Figure 2** *Effects of deforestation, Amazon Basin, Brazil*

Impact of deforestation in Indonesia on the carbon cycle

In the 1960s, 80% of Indonesia was rainforest. Since then, rapid development has decimated the natural rainforest and widespread deforestation has endangered species and disrupted the lives of indigenous people – just under half of the original forest cover now remains.

The destruction of the rainforest has had significant impacts on water and carbon cycles, changing the magnitude of important stores and flows. Once exposed, peatlands are easily eroded by wind and rain, and increased decomposition releases CO_2 into the atmosphere. The rainforests are no longer a carbon sink, but are now a carbon source.

Burning is a quick way to clear land prior to development. Fires are often widespread (Figure **3**) and left to burn out of control. They release huge quantities of carbon that had been stored for hundreds of years.

In 1997–98, the global growth rate of CO_2 in the atmosphere reached the highest on record. Much of this increase was attributed to forest fires in Indonesia – more than 8 million hectares of land were burnt.

Scientists believe that the extra carbon these fires pumped into the atmosphere exceeded all the carbon that living things on Earth remove from the atmosphere in one year.

Key

○ High confidence fires (high level of confidence that fire was started for forest-clearing purposes)

Indonesia Primary Forest 2012

☐ Degraded

■ Intact

⬥ **Figure 3** *NASA data from Global Forest Watch shows fire activity in Indonesia 10–17 August 2015. Many fires occurred in primary forest.*

Sixty second summary

- Photosynthesis is enhanced by the lush vegetation growth, so this is an important reducer of atmospheric CO_2.
- Deforestation and associated developments reduce the role of the rainforest as a carbon sink and releases carbon into the atmosphere.
- Deforestation reduces decomposition and plant and animal respiration.
- Deforestation increases the carbon content of the soil.

Over to you

Draw **two** spider diagrams to identify the key features of the tropical rainforest carbon cycle and the impacts of deforestation.

Case Study

You need to know:

- how to apply the water balance equation to a river catchment
- impacts of recent catchment developments on the water cycle.

Student Book
pages 52–3

Characteristics of the Exe upper catchment

The River Exe flows for 82.7 km from its source in the hills of Exmoor, through Exeter, to the sea at Exmouth on the south coast of Devon (Figure **1**).

- Physical – its highest point of 514 m is on Exmoor; the landscape is flatter in the south.
- Geology – most of the catchment is underlain by impermeable rocks, predominantly Devonian sandstones.
- Land use – most land is agricultural grassland (67%), with woodland (15%) and arable farmland. There are moors and peat bogs (3%) on high ground.

Water balance

The water balance for the Exe catchment can be expressed as:

precipitation (1295 mm) =
evaporation +/− soil water storage (451 mm) + runoff (844 mm)

Rainfall is high. Over Exmoor much of it is absorbed by the peaty moorland soils, although runoff is high when the soils are saturated or where drainage ditches have been dug.

Runoff accounts for some 65% of the water balance. There are two main reasons for this high figure:

- the impermeable nature of the most of the bedrock reduces percolation and baseflow
- drainage ditches on Exmoor reduce the amount of soil water storage.

Recent developments affecting the water cycle

Wimbleball Reservoir

In 1979 the River Haddeo, was dammed to create Wimbleball Reservoir (Figure **1**) to supply water to Exeter and parts of East Devon. It regulates water flow, preventing the peaks and troughs that make flooding or drought more likely.

Peatland restoration on Exmoor

For decades, drainage ditches have been dug in the peat bogs of Exmoor to make it suitable for farming. This has increased the speed of water flow to the River Exe. It also reduces water quality as more silt is carried downstream. As the peat has dried out, decomposition has occurred, releasing carbon to the atmosphere.

The Exmoor Mires Project works to restore the peat bogs (mires) by blocking the drainage ditches with peat blocks or moorland bales. This increases water content and returns the ground to the natural, saturated, boggy conditions, helping to retain carbon within the peat (see 1.18 and 6.12).

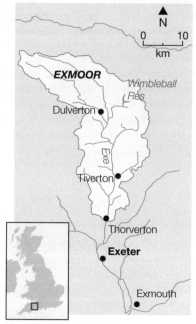

▲ **Figure 1** *The River Exe upper catchment, Devon*

 Sixty second summary

- Runoff is high (65%) because of impermeable rocks, steep valley slopes and the construction of drainage ditches.
- Wimbleball Reservoir is an important water store and has regulated water flow in the River Exe.
- Restoring peatland by blocking drainage ditches has increased the storage of water and raised water tables.

Over to you

Create a table outlining the main features of the River Exe catchment and the impacts of developments on the water cycle.

Student Book
pages 54–5

You need to know:

- how the Exmoor Mires Project impacts the River Exe water/carbon cycle
- that data can be used to investigate the effectiveness of peatland restoration.

The Exmoor Mires Project

The Exmoor Mires Project aims to restore 2000 ha of Exmoor to the boggy conditions that would naturally be present, by blocking drainage ditches with peat blocks and moorland bales (Figure **1**).

There are several benefits:

- More water storage in upper catchments – ensures a steady supply of water throughout the year.
- Improved water quality – slower throughflow means that less sediment is carried into the rivers.
- More carbon storage – by re-wetting peat and encouraging peat growth, CO_2 is naturally absorbed from the atmosphere and stored.
- Improved opportunities for education, leisure and recreation.
- Improved grazing and water supply for animals.
- Increased biodiversity.

By 2015, over 1000 ha of peat moorland had been restored and nearly 100 km of ditches blocked, raising the water table by 2.65 cm. Stormflow and flood peaks have been reduced.

Data collection

To measure the effectiveness of the Exmoor Mires Project on water tables and water transfers, dipwell *transects* (surveys along a line) were installed across newly blocked ditches. A dipmeter records the depth of the water table. This is inserted into the dipwell and when the electrodes make contact with water, a buzzer or light is activated and a depth measurement can be taken.

The conclusion was that the project was successful in increasing the capacity of the peat to store water, raising the water table and reducing flood peaks. This was consistent with other similar projects to re-wet peatlands in the UK.

Pages 54–5 in the student book has data relating to the Exmoor Mires project.

Figure 1 *Blocking ditches on Exmoor to increase storage capacity and reduce rates of throughflow*

 Sixty second summary

- The Exmoor Mires Project aims to restore peatlands to encourage increased water storage, regulate flows and raise water tables.
- The project will increase the role of the area as a carbon store/sink and promote wildlife.
- Blocking drainage ditches makes throughflow slower and increases storage capacity.
- The use of dipwell transects in designated areas has enabled changes to the water table to be monitored following the blockage of drainage ditches.
- Evidence indicates that water tables have risen and flood peaks in discharge have been reduced.

 Over to you

List the impacts of the Exmoor Mires Project on the water and carbon cycle.

2 Hot desert systems and landscapes

Your exam

(AL) *Hot desert systems and landscapes* is an **optional topic**. You must answer **one** question in Section B of Paper 1: Physical geography, from a choice of **three**: *Hot desert systems and landscapes* or *Coastal systems and landscapes* or *Glacial systems and landscapes*.

Paper 1 carries 120 marks and makes up 40% of your A Level. Section B carries 36 marks.

Specification subject content (Specification reference in brackets)

Either tick these boxes as a record of your revision, or use them to identify your strengths and weaknesses

Section in student book and revision guide	☹	😐	☺	Key terms you need to understand Complete the **key terms** (not just the words in bold) as your revision progresses. 2.1 has been started for you.
Deserts as natural systems *(3.1.2.1)*				
2.1 An introduction to hot deserts				*desert, aridity, distribution*
2.2 The characteristics of hot desert environments and their margins				
2.3 An introduction to the causes of aridity				
2.4 The hot desert water balance and the aridity index				
Systems and processes *(3.1.2.2)*				
2.5 Energy sources and sediment in hot desert environments				
2.6 Weathering processes in hot desert environments				
2.7 The role of wind in hot desert environments				

Arid landscape development in contrasting settings *(3.1.2.3)*				
2.8 Aeolian landforms in hot desert environments				
2.9 The role of water in hot desert environments				
2.10 Water-formed landforms in hot desert environments				
2.11 The development of characteristic desert landscapes				
Desertification *(3.1.2.4)*				
2.12 The changing extent of hot deserts				
2.13 Desertification – areas at risk and causes				
2.14 Impacts of desertification on ecosystems, landscapes and populations				
Case studies *(3.1.2.6)*				
2.15 Desertification in the Badia, Jordan				
2.16 The Mojave Desert, USA				

Note: The 'missing' specification reference **3.1.2.5** refers to skills

You need to know:
- the definition and distribution of hot deserts
- the elements comprising the hot desert system.

Student Book
pages 60–1

What is a hot desert?

A *hot desert* is a dry, barren region with little vegetation cover and very high temperatures. Deserts have less than 250 mm of precipitation a year, most of it in short, intense storms.

> **Big idea**
>
> Hot deserts landscapes result from the interaction of past and present-day processes at different scales and rates.

Where are hot deserts located?

Many of the world's hot deserts are found close to the Tropics of Cancer and Capricorn. Others are found in continental interiors (e.g. Africa and Australia) or close to mountain ranges (e.g. the Andes in South America). Hot deserts cover 25–30% of the Earth's land surface.

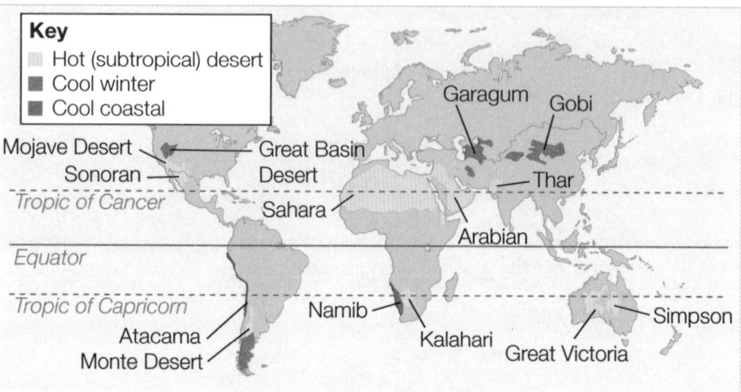

Key
- Hot (subtropical) desert
- Cool winter
- Cool coastal

Mojave Desert, Sonoran, Great Basin Desert, Garagum, Gobi, Thar, Tropic of Cancer, Sahara, Arabian, Equator, Tropic of Capricorn, Namib, Simpson, Atacama, Monte Desert, Kalahari, Great Victoria

Figure 1 ◗

Notice that the distribution of the world's deserts are closely aligned to the tropics

Desert landscapes

There are three main types of desert landscape:

- *Hamada* – bare rocky surfaces such as plateaus
- *Reg* – a stony desert where rock fragments are scattered over a large plain
- *Erg* – predominantly sand.

Desert landscapes result from complex interactions between geological structures, and weathering, mass movement, wind and water. Landforms develop over long periods of time, during which the landscape-forming processes will change.

A desert system is an open system with inputs, outputs, stores and transfers (Figure **2**).

Systems term	Hot desert
Input	Precipitation, solar radiation, descending air at the ITCZ.
Output	Runoff, reradiation of longwave radiation from the Earth's surface into the atmosphere, evaporation.
Energy	Latent heat associated with changes in the state of water. Energy associated with flowing water and moving air.
Stores/components	Playas (salt lakes), sand dunes.
Flows/transfers	Wind-blown sand, surface runoff, salinisation, sediment transfer.
Positive feedback loop	The presence or absence of vegetation can affect the regional climate. Less vegetation reduces the moisture emitted into the atmosphere; reduced humidity may lead to less rainfall and less vegetation.
Negative feedback	Intense weathering of a slope leads to a build-up of scree. If not eroded, this scree extends up the slopes, giving protection from weathering.
Dynamic equilibrium	Seasonal winds can lead to small-scale, short-term adjustments in sand dune profiles. But, over time, their shape remains broadly the same.

▲ **Figure 2** *Elements of a hot desert system*

 Sixty second summary

- Deserts are regions with average annual precipitation of below 250 mm a year.
- Most of the world's hot deserts are located close to the Tropics of Cancer and Capricorn.
- A hot desert landscape can be considered an open system, with inputs, components, processes and outputs.

 Over to you

Produce a spider diagram that summarises the different elements of a hot desert system.

You need to know:

- the characteristics of hot desert climates, soils and vegetation.

Student Book
pages 62–3

Desert climates

Figure **1** shows climatic conditions that are representative of the vast Saharan desert. Deserts exhibit a number climatic characteristics:

- Deserts are located in zones of high atmospheric pressure, where the air is sinking and becoming warm and dry.
- The lack of cloud cover results in high levels of sunshine and large diurnal (daily) temperature variations of up to 30 °C.
- Strong winds and sandstorms.
- Intense convective activity which can trigger thunderstorms.
- Some coastal deserts can have sea fog.

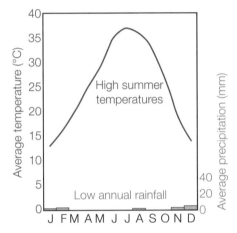

⚫ *Figure 1 This climate graph is typical of hot deserts*

Desert soils

The climate is too dry to support extensive vegetation so soils tend to be poorly developed, dry and sandy, with little humus. The red colour of many hot desert soils (*red desert soils*), is because of oxidation (where oxygen reacts chemically with iron-rich rocks). These soils are both porous and permeable with poor moisture retention.

Desert vegetation

The low rainfall and high temperatures and rates of evaporation make deserts hostile environments for plants, which require special adaptations.

- Drought avoidance – some plants only live for one season. They die during drought but store moisture, oil, fat, sugar and protein in seeds. Other plants use a long taproot to seek water deep underground.
- Drought resistance – some plants are able to adjust their metabolism, and become dormant. Others drop leaves to conserve moisture.
- Water storage – succulent plants (e.g. cacti), have shallow roots that quickly absorb water during a storm. Water is then stored in fleshy leaves, stems and roots.

Plants that have adapted to living in desert environments are called *xerophytes* (Figure **2**).

Plants can also have an impact on soil development and can create microclimates.

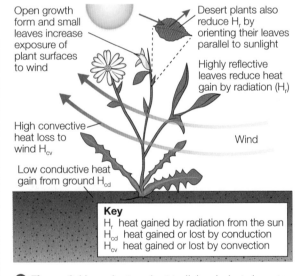

⚫ *Figure 2 How plants adapt to living in hot deserts*

⏱ **Sixty second summary**

- Desert climates typically have low rainfall (below 250 mm a year), very high summer temperatures and amounts of sunshine, and high diurnal temperature ranges.
- Soils are typically dry and sandy, with little organic matter due to the lack of (rotting) vegetation.
- Plant adaptations include small spiny leaves to reduce water loss, long taproots to seek water and fleshy stems to store water.

✏ **Over to you**

Make sure you can outline the key features of hot desert climates, soils and vegetation.

You need to know:
- the role of atmospheric processes, continentality and relief in the formation of hot deserts
- the effect of cold ocean currents on climatic processes and the formation of deserts.

Student Book
pages 64–5

What causes aridity and the formation of deserts?

Most of the world's hot deserts are found in a broad, but discontinuous belt between 20 and 25 degrees latitude. There are several factors that account for this distribution.

The global atmospheric circulation system

① Air rising in the ITCZ diverges both north and south to two circulation cells (Hadley cells).

② The air moving poleward cools and converges with the Ferrel cells, causing it to sink.

③ The sinking air becomes warmer and drier, forming two high pressure belts roughly 30° north and south of the Equator – the subtropical high.

High pressure (anticyclones) results in cloudless skies and lack of rainfall. The lack of cloud accounts for the high sunshine totals and the extremes of temperature.

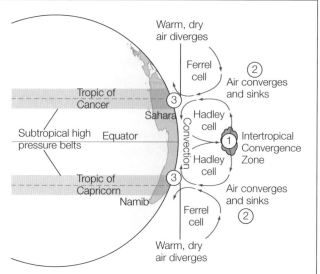

▲ **Figure 1** *The role of atmospheric processes in forming deserts*

Relief

Figure **2** shows that the *relief* of an area (and also cold ocean currents) can have an impact on the formation of deserts. This example is for the Andes Mountains and the Atacama Desert in South America.

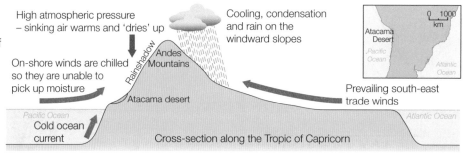

▲ **Figure 2** *Notice the importance of relief and cold ocean currents in forming the Atacama desert*

Continentality

Continentality is the way that a large land mass affects weather and climate. Aridity tends to be higher in continental interiors because the influence of moist airstreams from the oceans is reduced.

Cold ocean currents

Some deserts occur on coastlines adjacent to cold ocean currents, e.g. the Atacama Desert in South America. The air above the cold currents is cooled and sinks, remaining close to the ground. Overall precipitation is small because of the air's tendency to sink, despite the fog and drizzle often associated with the cold and moist sea air.

 Sixty second summary

- Atmospheric processes (in particular the global atmospheric system) create zones of aridity associated with subsiding air (high pressure belts) at the tropics.
- Aridity is high in continental interiors away from the influence of maritime processes.
- Relief can create a rainshadow effect on the leeward side of a mountain range, which can exacerbate aridity.
- Some deserts close to coastlines are affected by cold ocean currents due to the presence of high pressure (cold descending air) e.g. Atacama Desert.

 Over to you

Produce flashcards to record information about the factors affecting aridity (atmospheric processes, continentality, relief and cold ocean currents).

Student Book
pages 66–7

You need to know:
- what makes the hot desert water cycle distinctive
- how the aridity index is calculated and used to identify deserts
- the characteristics of the hot desert water balance.

What makes the hot desert water cycle distinctive?

Figure **1** shows the operation of the water cycle in a hot desert environment.
Its main features are:

① Rainfall is low and sporadic, often as heavy downpours that trigger overland flow.

② Actual evapotranspiration is low (low annual precipitation), but *potential evapotranspiration* is high.

③ Water from aquifers can reach the ground naturally through springs or be abstracted using wells.

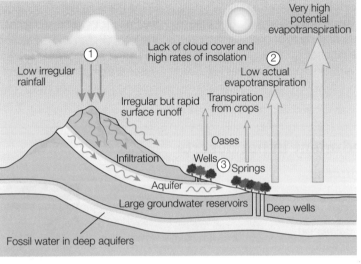

▲ **Figure 1** *The desert water cycle*

What is the aridity index?

The UNESCO **aridity index** is the ratio between mean annual precipitation (P) and mean annual potential evapotranspiration (PET).

Arid regions have a P/PET ratio of less than 0.20. This means that rainfall supplies less than 20% of the amount of water needed to support optimum plant growth. Semi-arid regions have an aridity index of 0.20–0.50 (20–50%).

What is the hot desert water balance?

The water balance shows the relationship between precipitation, soil moisture storage, evapotranspiration and runoff. It can be expressed as:

precipitation = evaporation +/– soil water storage + runoff

When PET exceeds precipitation, as it does in many hot deserts, there is no surplus water (runoff) for agriculture, industry or domestic consumption. The knock-on effect of this *water deficit* is that water has to be transferred from elsewhere or abstracted from fossil aquifers, which may be unsustainable.

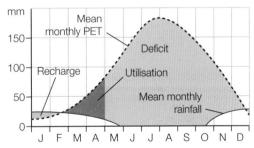

◀ **Figure 2**
Water balance graph for Baghdad, which is inland and has an annual rainfall of just 125 mm. Notice that for most of the year there is a significant soil moisture deficit

Sixty second summary

- The hot desert water cycle is characterised by low rainfall, high rates of potential evapotranspiration, variable river flows and important sources of groundwater.
- The aridity index is the ratio between mean annual precipitation and mean annual potential evapotranspiration.
- The hot desert water balance typically has a long period of severe water deficit and brief periods of soil water recharge and soil water utilisation.

 **Over to you**

Use a simple annotated sketch to summarise the key features of the hot desert water cycle and water budget.

Student Book
pages 68–9

You need to know:

- the sources of both energy and sediment in hot deserts
- characteristic features of sediment cells and sediment budgets.

What are the sources of energy in hot deserts?

Insolation is the amount of heat that reaches the Earth's surface.

- Temperature extremes can lead to insolation weathering such as **exfoliation** (see 2.6).
- The hot desert air is responsible for high rates of potential evapotranspiration.
- Variable warming causes localised winds which can shape landforms.

Water is scarce but, following a storm, the power of this water is immense with high rates of erosion in a short period of time.

Some rivers have their sources outside the deserts in wetter climates (e.g. the Nile in Egypt and the Euphrates in Iraq), and flow continuously through deserts, transporting sediment and creating fluvial landforms.

Wind is an important secondary source of energy in hot deserts. About 20–25% of desert surfaces are composed of sand and smaller-sized sediment, which can be moulded or transported by the wind.

The Harmattan wind blows from the Sahara Desert over West Africa between November and April creating powerful dust storms.

Sediment in hot deserts

Sources

Figure **1** shows several sources of sediment in hot deserts. Some sources are internal (green) and others may be external (red).

Mass movement (rockfalls)

Sediment washed onto the desert plain via gullies and wadis

Sand and finer sediments carried into the desert by winds or washed in by rivers from nearby mountains

Weathering of exposed rocks (exfoliation, chemical weathering)

Mass movement (soil creep, talus creep)

Erosion of rocks exposed on the flat desert plain

🔺 *Figure 1* Sediment sources in Monument Valley, Arizona, USA

Sediment cells

The movement of sediment can be described as a **sediment cell** with the following components:

- *inputs* – the sources of sediment (Figure **1**)
- *transfers* – movement by wind and water
- *sediment sinks* – areas of deposition
- *outputs* – transportation by wind or water.

Sediment budgets

A **sediment budget** considers the relative amounts of sediment in the hot desert system. Figure **2** suggests that surface erosion by slopewash was the main process of erosion. Only 22% of the total eroded material was deposited.

	Estimated average rates (tonnes/km²/yr)	% of total erosion
Total erosion	**5452.1**	**100**
Surface erosion (slopewash)	5335.2	97.9
Gully erosion	78.5	1.4
Mass movement	38.4	0.7
Total deposition	**1205.5**	**22**
Deposition in channels	564.9	10
Trapped in reservoir	640.6	12

🔺 *Figure 2* Sediment budget for the desert at Santa Fe, New Mexico, USA

Sixty second summary

- Insolation, water and wind are the three major sources of energy.
- Water and wind are powerful forces of landscape change (erosion, transport and deposition).
- Sources of sediment include weathering, mass movement and transfer by water/wind.
- A sediment cell is the movement of sediments, with inputs, transfers, sinks (stores) and outputs.
- Studies suggest that water is a major process of erosion.

Over to you

Make a list of the sources of both energy and sediment in hot desert environments. Explain the significance of **each** one.

You need to know:

- the processes and impacts of mechanical and chemical weathering.

Student Book
pages 70–1

Mechanical (physical) weathering

Thermal fracture

Read about the Peltier model of weathering on page 71 of the student book.

Thermal fracture is perhaps the most important type of mechanical weathering. It is the result of intense temperature fluctuations causing expansion and contraction, probably promoted by the presence of moisture. Colour, rock type and geology all determine the process of disintegration (Figure **1**).

Granular disintegration
Coloured minerals in rock heat up at different rates, facilitating break-up.

Exfoliation
Repeated heating of the outer surface of a rock leads to the outer skin peeling away.

Block separation
Breaking up of rock along joints and bedding planes.

Shattering
Rocks that are not granular or jointed tend to shatter into angular fragments.

🔺 *Figure 1 Characteristics of thermal fracture processes*

Salt crystallisation

Rainwater dissolves salts in the soil and is drawn to the surface by capillary action. Evaporation leaves salt crystals on the ground surface. Over time, the growth of these crystals causes stresses that break up rocks.

Frost shattering

Water (often dew) freezes and expands in cracks and pores within rocks. Repeated freezing/thawing shatters rocks.

Chemical weathering

Rocks that change as a result of chemical action, usually in the presence of water. The chemical process is enhanced in high temperature. Rocks containing salts are vulnerable to being dissolved. This is *solution*. Once dissolved, the high rates of evaporation will lead to the deposition of salty deposits on the ground. Flaking or pitting of rock surfaces is evidence of chemical weathering in hot deserts.

🔺 *Figure 2 Exfoliation – the outer skin of the rock has been peeled away*

⏱ Sixty second summary

- Thermal fracture, salt crystallisation and frost shattering are processes of mechanical weathering.
- Solution is an important process of chemical weathering in hot deserts.
- Weathering is an important source of sediment in hot deserts, providing tools for water and wind erosion.
- Weathering produces sediment that can be transported and redeposited as, for example, alluvial fans and sand dunes.

✏ Over to you

Summarise the key features of the processes of weathering operating in hot deserts. Group them into 'mechanical' and 'chemical' weathering.

You need to know:
- the processes of wind erosion and transportation
- the factors responsible for wind deposition.

Student Book
pages 72–3

Wind erosion

There are two main types of erosion by wind – deflation and abrasion.

Deflation is the removal of loose material from the desert floor, lowering the desert surface over time. This often results in the exposure of the underlying bedrock. Strong eddies or localised winds can hollow-out the desert surface to produce a *deflation hollow*, which can be extensive and cover thousands of square kilometres.

Abrasion is wind-blown sand, which carves or sculpts rock into a variety of shapes, usually within a metre of the desert floor (Figure 1). The strength, duration and direction of the wind affect the intensity of abrasion, as does the nature of the sand (rock type and its angularity) and the rock outcrops.

Wind deposition

Deposition of sand will take place when wind velocity falls below that required to transport the sand. This velocity will vary according to the size (mass) of the sand grains being transported.

Sand will be deposited in sheltered areas protected from the wind.

⬢ **Figure 1** *Mushroom rock carved by abrasion*

Wind transportation

Suspension – fine particles are picked up and carried considerable distances, sometimes beyond the margins of the desert itself.

Saltation – sand particles are picked up and carried a few centimetres by wind.

Surface creep – sand and pebbles are rolled along the desert floor.

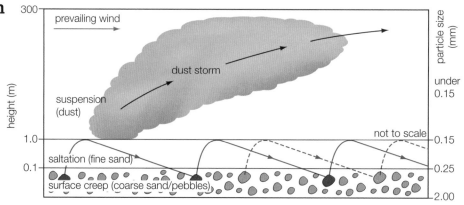

⬢ **Figure 2** *Processes of transportation; particle size and height above the ground affects the process of transportation*

⏱ **Sixty second summary**

- Wind erosion is dominated by deflation (the removal of loose material from the desert floor) and abrasion (sand blasting and sculpting of rocks by sand suspended in the air).
- Wind transportation involves the processes of surface creep (rolling on the desert floor), saltation (low bouncing) and suspension (held within the air).
- Wind deposition occurs when wind velocities fall, often in sheltered locations (e.g. the lee of a rock outcrop).

✎ **Over to you**

Write down the key processes and factors affecting wind erosion, transportation and deposition.

You need to know:
- the characteristics and formation of landforms resulting from aeolian (wind) processes.

Student Book
pages 74–7

Landforms resulting from aeolian (wind) processes

Wind is a potent source of energy in hot deserts and is responsible for the formation of a number of distinctive landforms.

Deflation hollows

Loose material is eroded by winds, and transferred and deposited elsewhere, creating a depression in the desert surface – a *deflation hollow*.

This is an example of the desert system in action – landscape change by the flow or transfer of energy. The largest in the world is the Qattara Depression in Egypt which is now 134 m below sea level.

Desert pavements

A **desert pavement**, or *reg*, is a desert surface covered with rock fragments often resembling a cobbled pavement. This rocky pavement is formed when the wind blows away the finer sands, leaving behind the larger stones.

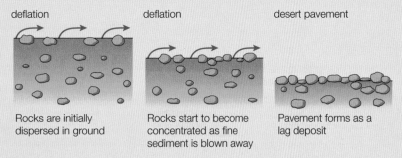

deflation   deflation   desert pavement

Rocks are initially dispersed in ground | Rocks start to become concentrated as fine sediment is blown away | Pavement forms as a lag deposit

🔺 *Figure 1 Formation of desert pavement. Notice that stones are lowered as the underlying material is eroded.*

Ventifacts

Ventifacts are individual rocks, usually the size of pebbles that have a clearly eroded face (*facet*) that is aligned with the prevailing wind.

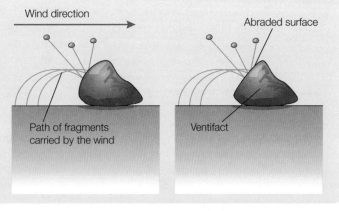

Wind direction

Abraded surface

Path of fragments carried by the wind

Ventifact

Figure 2 ◗
Wind abrasion creates an angular facet on the ventifact

Yardangs

Resembling the hull of upturned ships, **yardangs** are elongated ridges separated by deep grooves cut into the desert surface.

The weaker rocks are eroded by abrasion to form deep troughs, leaving the tougher rocks upstanding – it is these more resistant rocks that form the yardangs. Winds need to blow in just one direction only for them to form. Size can vary from just a few centimetres in height and length to hundreds of metres in height and several kilometres in length.

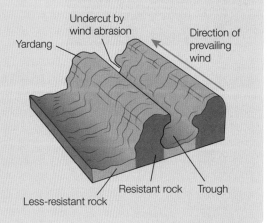

Undercut by wind abrasion

Direction of prevailing wind

Yardang

Resistant rock   Trough

Less-resistant rock

Figure 3 ◗
The formation of yardangs with vertical layers of rock parallel to the prevailing wind

Continued over >>>

Zeugen

As with yardangs, **zeugen** also form ridges, in some cases, up to 30 m high. The key difference is that zeugen develop in horizontally layered rocks (rather than in rocks that are layered vertically). This gives them a pedestal-like shape, with a flat-topped 'cap rock' protecting the less-resistant underlying layers.

The primary process operating on zeugen is abrasion. With most abrasion being concentrated within a metre or so of the desert surface, they often have a slightly narrower, more eroded lower portion.

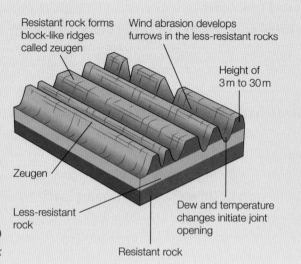

Figure 4 ▸

The formation of zeugen with horizontal layers of rock

Sand dunes

There are two common forms of sand dune: barchans and seif dunes.

* **Barchans** are crescent-shaped sand dunes, often found in isolation in deserts where there is a relatively limited supply of sand but a dominant wind direction. They form at right angles to the prevailing winds (Figure **5**).

* **Seif dunes** are elongated linear sand dunes that are commonly found in extensive areas of sand called sand seas. They can stretch for several hundred metres and are formed parallel to the prevailing wind direction. Seif dunes may develop from barchans.

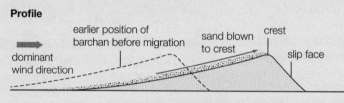

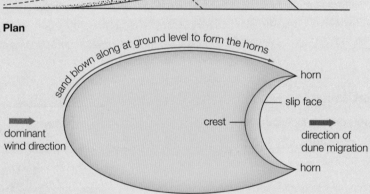

 Figure 5 *This diagram shows the characteristic features and formation of barchans*

Figure 13 on page 77 of the student book shows the formation of a seif dune.

Figure 13 on page 77 of the student book shows the formation of a seif dune.

Sixty second summary

* Aeolian processes are important in forming distinctive desert landforms.
* Deflation hollows, formed by the erosion of loose sediment on the desert floor, can cover vast areas.
* Desert pavements are left behind when finer sediment is removed.
* Individual rocks can be sculpted by sand to form angular ventifacts.
* Geology determines the shape and characteristic features of yardangs and zeugen.
* Sand dunes can be extensive. They can be broadly split into crescent-shaped barchans and longitudinal seif dunes.

Over to you

Use simple annotated sketches to summarise the distinctive characteristics and formation of **three** aeolian (wind) landforms in hot deserts.

You need to know:

- the sources of water in hot deserts
- the processes of erosion, transportation and deposition in hot deserts.

Student Book
pages 78–9

The role and sources of water in hot deserts

In mountain ranges, erosion by water is the dominant process following storm events or periods of snowmelt. On lowland desert plains, deposition is the main process.

There are four main sources of water.

- **Exogenous rivers** have their source outside desert regions. They have sufficient water to flow continuously despite the high rates of evaporation (e.g. the Colorado, the Indus and the Nile).
- **Endoreic rivers** flow into deserts but usually terminate in a lake or inland sea (e.g. the River Jordan drains into the Dead Sea).
- **Ephemeral rivers** flow intermittently after a storm event or snowmelt. In times of flood they can be powerful forces of erosion.
- **Episodic flash floods** tend to result from convectional storms unleashing large amounts of water in a short period of time. Water flows as a sheet (**sheet flooding**) or is confined within a channel (**channel flash flooding**). Huge amounts of sediment can be washed out of the mountains to be deposited as alluvial plains on the lowlands below.

 Big idea

Despite its relative scarcity, water is believed to be a dominant force in hot desert landscape development.

 Figure 1 Alluvial fan, Death Valley, USA

Water action processes in hot deserts

Erosion
• Hydraulic action – the sheer force of the water
• Abrasion – sandpaper effect of loose rocks being ground over bedrock
• Corrasion – fragments of rock carried by water gouging or sculpting bedrock
• Solution – dissolving of soluble rocks, such as limestone

Transportation
• Traction – rolling sediment along the surface or channel bed
• Saltation – bouncing or leaping motion of particles
• Suspension – sediment carried within the body of water
• Solution – dissolved sediment

Deposition
• Takes place where velocity drops, such as on the inside of meander bends or where streams flow out of mountain edges onto flat desert plains (alluvial fan)

 Figure 2 Summary of river processes

Splash erosion	• The force of falling rainwater displacing soil particles.
Sheet erosion	• Water running as a sheet over impermeable surfaces or compacted soil washing away disturbed particles.
Rill erosion	• Sheetwash wears down the soil to form a definite path to form rivulets in the soil, called rills.
Gully erosion	• Over time, rills become wider and deeper to form gullies.
Bank erosion	• Fast water flow wears away the stream sides, causing the banks to collapse and the channel to widen.

 Figure 3 Summary of sheetwash processes

See page 79 of the student book for a case study of an episodic flash flood in Morocco.

Sixty second summary

- Water action is a dominant geomorphological influence in most of the world's deserts.
- In mountain ranges, erosion dominates following storms or snowmelt, whereas deposition is the main process in lowland areas.
- Exogenous rivers have their source outside desert regions and retain enough water to flow through deserts.
- Endoreic rivers flow into deserts where they terminate, usually in a salt lake.
- Ephemeral rivers flow intermittently following a rain storm or seasonal snowmelt.
- Episodic flash floods occur after torrential rain events and may involve sheetflow or channel flash flooding.

Over to you

Create simple sketches with annotation to summarise the **four** different sources of water in hot deserts.

You need to know:

- the distinctive characteristics and formation of landforms in a hot desert environment.

Student Book
pages 80–3

Landforms resulting from water processes

Water erosion, transportation and deposition play important roles in the formation and shaping of many desert landforms, and in creating distinctive hot desert landscapes (Figure **1**).

Some of these features are still being actively formed by erosion (e.g. wadis and canyons). Others are gradually being diminished by a combination of erosion, weathering and mass movement (e.g. mesas and buttes).

Figure 1 ➤
Landforms created by water action in hot deserts

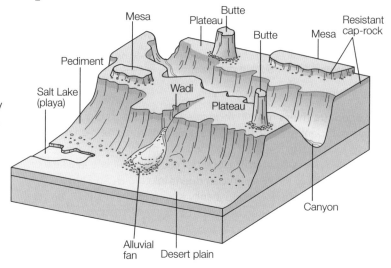

Wadis

A **wadi** is a distinct dry riverbed in a lowland plain or an incised gully/valley cut into a plateau. They can be just a few metres or many hundreds of kilometres in length. Common features include:

- steep edges due to severe erosion during periods of high water flow
- flat bottoms infilled with coarse sediment as the finer material is washed out or removed by wind action
- *a braided* drainage pattern due to the high quantities of transported sediment and the 'flashy' nature of the water flow.

Alluvial fans and bajada

At the edge of a mountain range, sediment washed out through a wadi or canyon is deposited to form an **alluvial fan**.

- As the river water spreads from the mountain front, energy is lost and sediment is deposited.
- This sediment is layered and sorted by the water.
- Coarse sediment is deposited closest to the mountain range; finer sediment is washed further away.
- Alluvial fans can extend for several kilometres and reach thicknesses of up to 300 m.
- Alluvial fans can merge to form a continuous apron of sediment called a **bajada**.
- Playas (salt lakes) may form on lower bajadas.

Pediments

A **pediment** is a gently-sloping erosional rock surface at the foot of a mountain range. A common characteristic is a distinct change in angle between the mountain front and the top of the pediment. Sediment transported by rivers from nearby mountains, often covers pediments (e.g. alluvial fans). Periodic sheetwash then removes the loose material creating a smooth rock surface.

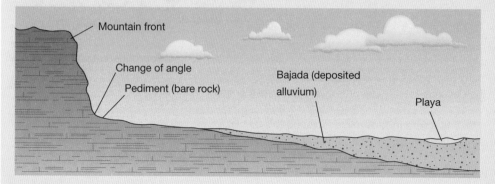

◀ **Figure 2**
Pediments and bajadas both form at the foot of a mountain front; note the difference between the two

Playas (salt lakes)

Water on the desert floor will percolate through the coarse sediment and be transferred as throughflow, to collect in a large hollow where it may form a **playa** – an enclosed desert lake with no outflow. Playas are shallow with a large surface area. This encourages evaporation and a salty crust will often form around their edges. In some deserts, the accumulation of salt is sufficient for it to be exploited commercially (e.g. the Chott el Djerid in southern Tunisia).

Mesas, buttes and inselbergs

These are relic features that have become isolated by river erosion (Figure **3**).

- **Mesas** are large plateau-like features often bordered by steep wadis or canyons.
- **Buttes** are smaller pinnacles of rock at a more advanced stage of development.
- **Inselbergs** develop in rocks with no layering or with variable rock strengths (e.g. granite or sandstone). They were possibly created in climatic conditions where chemical weathering was more active, creating more rounded structures.

Canyons

Canyons are steep-sided gorges mostly associated with rapid downcutting into plateaus. Some very deep canyons are formed by exogenous rivers that flow permanently through a desert region (e.g. the Grand Canyon in Arizona formed by the Colorado River).

Figure 3 *Mesas and buttes in Monument Valley, Arizona, USA*

 Sixty second summary

- Water erosion, transportation and deposition are responsible for a wide variety of landforms in hot deserts.
- Wadis are steep-sided dry valleys formed by intense erosion during periods of high water flow.
- Alluvial fans result from the deposition of sediment at the foot of mountains. Several can merge to form a bajada.
- A pediment is a gently sloping rocky surfaces formed by sheetwash erosion at the edge of a mountain range.
- A playa is a shallow salt lake at the terminus of surface streams or where groundwater collects in a depression.
- Canyons are steep-sided gorges, often with rivers flowing through them.
- Mesas, buttes and inselbergs are relic features of a mountain landscape.

 Over to you

From memory, and using sketches if necessary, outline the key characteristics and formation of water-formed landforms in hot desert environments.

You need to know:

- the role of present and past conditions in determining desert landscapes
- the concept of 'basin and range' topography associated with tectonic activity
- how models are used in describing cycles of erosion.

The development of desert landscapes

Every desert landscape is unique because of the interaction between the rocks and the processes operating on them, both past and present.

These processes are essentially water and wind action, although weathering, mass movement and vegetation are also important drivers in the desert system.

The importance of time

Other than during a sandstorm or following a period of heavy rain, processes cannot be seen operating. It can therefore be difficult to make connections between processes and landforms, especially as they may have formed in different climatic conditions than exist today.

Figure **1** shows a typical 'basin and range' landscape associated with tectonic activity where vertical displacement has taken place along a fault. This landscape exhibits some of the features studied earlier. It demonstrates the complexity of a landscape affected by a range of different processes over a long period of time.

  Big idea

Geomorphology is essentially a study of landform evolution resulting from processes acting over long periods. Think, therefore, of a desert landscape as a snapshot in time.

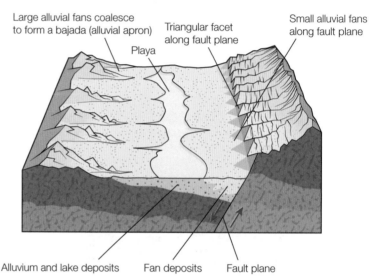

Large alluvial fans coalesce to form a bajada (alluvial apron) — Triangular facet along fault plane — Small alluvial fans along fault plane — Playa — Alluvium and lake deposits — Fan deposits — Fault plane

Figure 1 *The effect of faulting in creating a basin and range topography*

King's arid cycle of erosion

In the 1940s the desert geomorphologist L.C. King developed his *arid cycle of erosion*. Although highly simplified, it makes the point that landforms are dynamic and evolving elements of a changing landscape. It also stresses the importance of time in understanding the development of desert landscapes.

W.M. Davis developed his own model of how landscapes change over time. Both King's and Davis' concepts are at the heart of modern-day study in that the systems approach enables a causal understanding to be gained of the links between process and landforms.

See Figure 3 on page 85 of the student book for a diagram of King's *arid cycle of erosion.*

 Sixty second summary

- A desert landscape reflects the interaction of geology and processes operating both in the present and the past.
- 'Basin and range' topography develops in landscapes affected by tectonic activity where faulting has occurred.
- Models have been developed (e.g. King and Davis) to describe the development of desert landscapes often invoking the concept of a 'cycle' of development.

 Over to you

Outline the key features associated with 'basin and range' topography and to describe King's arid cycle of erosion – consider using a flow diagram.

You need to know:

- how climatic conditions in the past have affected the extent and distribution of the world's deserts.

Student Book
pages 86–7

How has the extent and distribution of deserts changed?

The present-day extent and distribution of deserts has remained almost unchanged for the last 5000 years or so. This reflects a largely stable climate with few major shifts or changes.

However, between the maximum extent of the last glacial period and 5000 years ago, there were considerable climatic fluctuations triggering dramatic changes in the extent and distribution of the world's deserts.

Figure **1** shows how the extent and distribution of deserts has changed over the last 20000 years

The ever-changing Sahara Desert

It is hard to believe that just a few thousand years ago much of the Sahara Desert was covered in grass and parts of it were even swampy!

Scientists believe that a burst of monsoon rains transformed the desert into grasslands over a period of just a few hundred years. This lasted until about 7500 years ago when conditions started to become more arid and less hospitable for human settlement.

From about 5000 years ago, conditions became more or less as they are today.

Last glacial maximum (about 20000 years ago)

During the last glacial maximum, aridity was very widespread. Deserts in the far north were cold deserts. Further south, deserts existed in similar locations to the present day but they were far more extensive.

About 8000 years ago

During this interglacial period conditions were very much warmer and more humid. Forests were widespread, thriving in warm and wet conditions. Aridity fell dramatically, with many of the present-day deserts becoming grasslands.

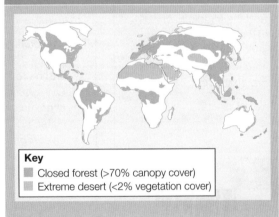

5000 years ago – present

The monsoon rains over Asia and Africa began to diminish. This was the start of conditions becoming more arid. By 3000 years ago, conditions were much as they are today. This map shows the 'present potential' vegetation cover and does not take account of the fact that large areas of forest have been cleared by people to make way for farming.

Key
- ■ Closed forest (>70% canopy cover)
- ■ Extreme desert (<2% vegetation cover)

▲ **Figure 1** *Notice the changes in location and extent of the world's deserts*

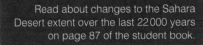
Read about changes to the Sahara Desert extent over the last 22000 years on page 87 of the student book.

Sixty second summary

- The climate has changed significantly in the 10000 years since the last glacial period.
- Changes in global climates have affected the extent and distribution of hot deserts.
- Aridity was widespread during the last glacial maximum.
- The post-glacial period was wetter during which most of the current hot deserts became grasslands.
- Deserts have only existed in their present extent and distribution for the last 5000 years.

 Over to you

Create a timeline to summarise the changing extent and distribution of deserts over the last 10000 years.

You need to know:

- what desertification is
- how climate change can be considered to be a 'natural' cause of desertification
- how people can contribute to desertification through poor land management.

Student Book
pages 88–9

What is desertification?

Desertification is *'the destruction of the biological potential of the land, which can lead ultimately to desert-like conditions'.*

Land that was once marginal is turned into an unproductive wasteland as ecosystems are destroyed, vegetation dies and soil becomes exposed and eroded.

Figure **1** shows those areas that are most at risk.

Figure 1 ➲

Land vulnerable to desertification. Notice that the most vulnerable areas are closest to existing deserts.

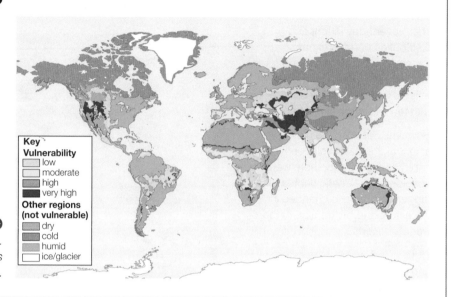

Key
Vulnerability
- low
- moderate
- high
- very high

Other regions (not vulnerable)
- dry
- cold
- humid
- ice/glacier

Natural causes of desertification

The climate has changed throughout time. About 8000 years ago North Africa and the Middle East were much wetter than today. Recent climate change has seen global temperatures increase by about 1 °C compared to 1880. In recent decades rainfall patterns (Figure **2**) and temperatures are changing.

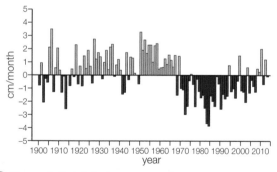

▲ **Figure 2** *Rainfall trends in the Sahel. Droughts have become more common since the 1970s, although there is a recent trend towards wetter periods.*

Human causes of desertification

People can contribute to the process of desertification on marginal land:

- *Overcultivation* – intensive farming reduces soil fertility. The lack of organic matter make soils friable and more likely to be washed or blown away.
- *Overgrazing* – overstocking exceeds the number of animals that the land can support and soils will suffer.
- *Overirrigation* – causes salts to be dissolved in the soil, which form a salty crust on the surface (salinisation). This creates impermeable, infertile soil.
- *Population increase* – puts pressure on land to be as productive as possible.
- *Firewood* – in semi-arid regions people rely on wood for cooking. Land becomes stripped of trees exposing the soil to wind and rain, leading to soil erosion.
- *Tourism* – safari minibuses can cause serious damage to vegetation, leading to soil erosion.

 Sixty second summary

- Desertification is a process whereby marginal land turns into desert.
- Climate change is a 'natural' cause, with rainfall becoming less predictable and droughts more common.
- Human actions, such as overgrazing, poor irrigation practices and firewood collection can inadvertently contribute to desertification.

 Over to you

Summarise the main factors contributing to desertification. Group them as 'Human' factors and 'Natural' factors.

You need to know:

- the impacts of desertification on ecosystems, people and landscape processes
- how future climate change may affect desertification.

Student Book
pages 90–1

What are the current impacts on ecosystems?

Desertification affects ecosystems in several ways:

- Overexploitation by agriculture reduces *biodiversity* (Figure **1**).
- Loss of topsoil due to increasing exposure to wind and rain.
- Increased salinity if irrigation is used.
- Water sources become depleted and plants and animals die.
- Vegetation destruction causes animals to migrate.

Figure 1 ▶

Links between desertification, biodiversity loss and climate change

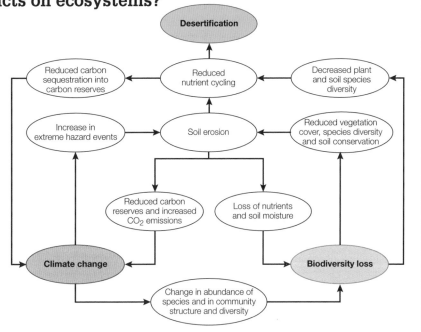

What are the current impacts on people?

Desertification affects people in several ways:

- Lack of water creates hygiene and disease issues.
- Loss of vegetation reduces food productivity and food security.
- Dust clouds affect air quality and health.
- Depleted grasslands cause animals to die or become emaciated, forcing population migration to already overcrowded cities.

What are the current impacts on landscapes?

Desertification affects the landscape in several ways:

- Reduced sand dune formation as less vegetation to trap loose material.
- Soil erosion by water causing deep gullies.
- Landslides on destabilised slopes where vegetation is removed.
- More frequent sandstorms as the protective layer of vegetation has been removed.

How might future climate change affect desertification?

By 2050 temperatures in the world's drylands could rise by 2 to 5 °C. This will increase desertification with the following possible scenarios:

- Less productive soils will lead to settlement and agriculture becoming less viable and many people will migrate.
- Migration could put pressure on already overcrowded cities.
- Some may stay and adapt by using soil conservation strategies or changing agricultural practices.

Sixty second summary

- Desertification can have serious impacts on ecosystems; habitats destroyed, biodiversity reduced, soils depleted of nutrients and vulnerable to increasing salinity.
- Unproductive land increases the risk of food insecurity.
- As vegetation dies, soils are also prone to wind and water erosion.
- Climate change may accelerate rates of desertification, causing some people to abandon settlements and migrate to already overcrowded cities.

Over to you

Write down **two** impacts of desertification on **each** of ecosystems, people and physical landscapes.

You need to know:
- the causes of desertification in the Badia
- how the process of desertification has been reversed.

Student Book
pages 92–3

Where is the Badia?

The Badia is a vast, sparsely populated desert region in eastern Jordan inhabited by traditional Bedouin who herd sheep, goats and camels. Annual rainfall is less than 150 mm (much as torrential storms) and temperatures can reach over 40°C in summer, and fall below freezing in winter.

What has caused desertification?

The Tal Rimah Rangelands near As Safawi have been grazed sustainably for hundreds of years.

- Following the 1991 Gulf War, large numbers of sheep were bought by farmers cheaply from Iraq.
- Overgrazing on the Rangelands desertified the land.
- The land was abandoned by farmers.

Addressing desertification

In 2002 the Tal Rimah Rangeland Rehabilitation Project initiated the construction of stone walls to control and retain water in order to encourage growth of vegetation – *water harvesting* (Figure **2**). Drought-tolerant shrubs have also been planted which:

- provide grazing for animals and fuel for firewood
- have roots that hold the soil together
- encourage species diversity by offering different habitats.

How successful has the project been?

Careful management resulted in:

- numbers of plant and animal species increasing from 21 to 54 between 2004 and 2008
- reintroduction of sheep by 2008
- flowering plants attracting butterflies and other insects
- birds nesting in the base of shrubs
- the return of other animals and reptiles.

Evaluating the human responses

Communities had to support the project in order for it to be a success. Local people had to show resilience and patience by resisting the temptation to return to the land too soon. They adapted their farming techniques and stock management to enable the new plants to become established and avoid overgrazing.

New shrubs and a greater variety of plants are now established and the future looks promising.

 Big idea

The process of desertification can be reversed with careful long-term management.

⬣ **Figure 1** *Location of Jordan and the Badia*

⬣ **Figure 2** *Stone walls retain topsoil and harvest water, encouraging vegetation growth*

Sixty second summary

- The Badia is a desert region inhabited by traditional Bedouin animal herders.
- Overgrazing after the 1991 Gulf War caused desertification and migration.
- Careful management and community support has reclaimed the area using stone walls and specially adapted plants.
- Animals have been reintroduced and biodiversity has increased.
- The Tal Rimah Rangeland Rehabilitation Project demonstrates that desertification can be addressed.

Over to you

Create a timeline to summarise the changes that have taken place in the Badia to address the problem of desertification.

Student Book
pages 94–7

You need to know:

- the location and characteristics of the Mojave Desert
- the distinctive landforms and landscapes of the Mojave Desert.

Case Study

Where is the Mojave Desert?

The Mojave Desert is located in south-west USA covering an area of 124 000 km².

It occupies parts of California, Nevada, Utah and Arizona. It has the following characteristics:

- Displays classic basin and range topography. High points are over 3000 m; Death Valley is 86 m below sea level.
- Annual rainfall of less than 140 mm; strong winds; summer temperatures can reach 50 °C; winter temperatures can be –7 °C.
- Exogenous rivers flow through the desert, including the Colorado River.
- Sparsely populated, but with a few large cities (e.g. Las Vegas).

Landforms and landscapes of the Mojave Desert

At the foot of the weathered granite mountains are alluvial fans which spread out over the desert plain, sometimes merging to form bajadas. Bare rock pediments are exposed in places. Playa lakes are common on the plains. In places, there are extensive, highly mobile sand dunes. There is evidence of wind and water being important in landscape development – the deeply eroded canyons of the mountains (water) provide sand for sand dunes.

Sixty second summary

- The Mojave Desert is located in the south-west of the USA.
- The high desert landscape is typical 'basin and range' landscape.
- Water action is particularly important in both sculpting the landscape and providing sand for the creation of sand dunes.
- There are many classic desert landforms including alluvial fan, pediment and inselbergs, playa and sand dunes.

Over to you

Name some landforms of the Mojave Desert. Think back to 2.8 and 2.10 to explain their formation.

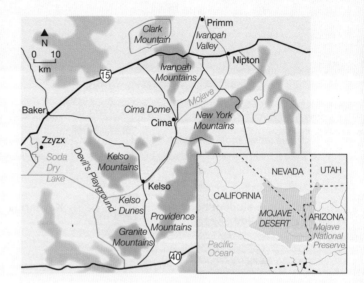

⊘ **Figure 1** *The location of the Mojave National Preserve*

The Mojave National Preserve

The Mojave National Preserve is a protected part of the Mojave Desert (Figure **1**), with several examples of classic desert landforms in and around the area.

Alluvial fan: Lucy Gray Fan radiates out from the Lucy Gray Mountains drains into the Ivanpah Valley. Coarse sediment is at the top of the fan while finer material is spread out over the desert plain.

Playa: the Mojave River brings clays and muds into the basin, and springs generate water for Soda Lake. Salt crusts develop in the summer which winds pick up creating a dusty haze in the air.

Pediment: Cima Dome is an ancient rocky pediment, left behind after erosion and weathering stripped away the mountainous landscape, leaving behind an extensive rocky pediment.

Sand dunes: Kelso Dunes comprise mobile and stabilised (partly vegetated) dunes. Most of the sand comes from the San Bernadino Mountains and is deposited by the Mojave River, and transported by the wind to form the dunes.

3 Coastal systems and landscapes

Your exam

(AL) *Coastal systems and landscapes* is an **optional topic.** You must answer **one** question in Section B of Paper 1: Physical geography, from a choice of **three:** *Hot desert systems and landscapes* **or** *Coastal systems and landscapes* **or** *Glacial systems and landscapes.*
Paper 1 carries 120 marks and makes up 40% of your A Level. Section B carries 36 marks.

(AS) *Coastal systems and landscapes* is an **optional topic.** You must answer **one** question in Section A of Paper 1: Physical geography and people and the environment, from a choice of **three:** *Water and carbon cycles* **or** *Coastal systems and landscapes* **or** *Glacial systems and landscapes.*
Paper 1 makes up 50% of your AS Level. Section A carries 40 marks.

Specification subject content
(Specification reference in brackets)

Either tick these boxes as a record of your revision
or use them to identify your strengths and weaknesses

Section in student book and revision guide	☹	😐	☺	Key terms you need to understand Complete the **key terms** (not just the words in bold) as your revision progresses. 3.1 has been started for you.
Coasts as natural systems *(3.1.3.1)*				
3.1 Coasts as natural systems				*open system, inputs, outputs,*
Systems and processes *(3.1.3.2)*				
3.2 Sources of energy at the coast				
3.3 Sediment sources, cells and budgets				
3.4 Weathering, mass movement and runoff				
3.5 Marine processes – erosion, transportation and deposition				

Coastal landscape development *(3.1.3.3)*			
3.6 Landforms and landscapes of coastal erosion			
3.7 Landforms and landscapes of coastal deposition			
3.8 Sea level change			
Coastal management *(3.1.3.4)*			
3.9 Coastal management			
Case studies *(3.1.3.6)*			
3.10 Coastal processes on the Holderness coast			
3.11 Risk and opportunity in Odisha, India			

Note: The 'missing' specification reference **3.1.3.5** refers to skills

You need to know:

- how the systems approach can be applied to the coast
- that the coastal system links to other natural systems.

Student Book
pages 102–5

The coast as an open system

The coast is an example of an open system. This means that it has inputs that originate from outside the system (such as sediment carried into the coastal zone by rivers) and outputs to other natural systems (such as eroded material transported to the ocean).

As an open system, the coast has important links with other natural systems such as the atmosphere (e.g. wind in generating waves), tectonics, ecosystems and oceanic systems. These natural systems are linked together by flows of energy and by the transfer of material.

 Big idea

The coast is an open system which is linked to other natural systems.

The coastal system

Systems terminology helps us to understand the connections between processes and landforms.

Systems term	Definition	Coastal example
Input	Material or energy moving into the system from outside	Precipitation, wind
Output	Material or energy moving from the system to the outside	Ocean currents, rip tides, sediment transfer, evaporation
Energy	Power or driving force	Energy associated with flowing water, the effects of gravity on cliffs and moving air (wind energy transferred to wave energy)
Stores/components	The individual elements or parts of a system	Beach, sand dunes, nearshore sediment
Flows/transfers	The links or relationships between the components	Wind-blown sand, mass movement processes, longshore drift
Positive feedback	Where a flow/transfer leads to increase or growth	Coastal management can inadvertently lead to an increase in erosion elsewhere along the coast. Groynes trap sediment, depriving areas further down-drift of beach replenishment and this can exacerbate erosion.
Negative feedback	Where a flow/transfer leads to decrease or decline	When the rate of weathering and mass movement exceeds the rate of cliff-foot erosion a scree slope is formed. Over time, this apron of material extends up the cliff face protecting the cliff face from subaerial processes.
Dynamic equilibrium	This represents a state of balance within a constantly changing system	Constructive waves build up a beach, making it steeper. This encourages the formation of destructive waves that plunge rather than surge. Redistribution of sediment offshore by destructive waves reduces the beach gradient which, in turn, encourages the waves to become more constructive.

▲ **Figure 1** *Definitions and examples of the aspects of the coastal system*

Sediment cells

A clear example of the application of systems concepts to the coast is the sediment cell. There are eleven major sediment cells in England and Wales, which form the basis for coastal management (Figure **2**). Here, there are clear inputs of sediment (e.g. from rivers and cliff erosion), transfers of sediment (e.g. longshore drift), stores (e.g. beaches and spits) and outputs (e.g. transfer to the deep ocean). (Also see 3.3.)

Figure 2 ◗

Sediment cells in England and Wales are largely self-contained but do have inputs to, and outputs from each cell

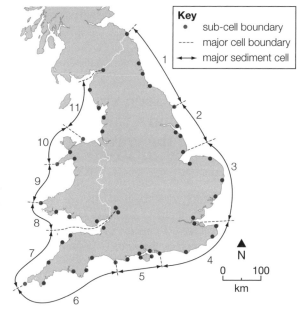

Key
- • sub-cell boundary
- ---- major cell boundary
- ←→ major sediment cell

Links between the coastal system and other natural systems

Coastal systems are interlinked with other physical and human systems. They both affect and are being affected by change – if one changes, they are all subject to change.

- During the Quaternary glacial and interglacial periods, sea levels rose and fell several times in response to changes in the global water cycle.
- This affected the location of coastal processes – some landforms owe their development to changes in the sea level (Figure **3**).

- Recent changes in the global carbon cycle, related to global warming, are affecting sea levels. This, in turn, is affecting coastal and terrestrial ecosystems.
- These changes impact on human systems with risk of severe flooding and storm surges at the coast.

Dyrhólaey, Iceland

Figure **3** shows the coast at Dyrhólaey, a small peninsula on the south coast of Iceland. The next land mass to the south is Antarctica so waves driven over this vast area of ocean are responsible for extremely active erosion and significant transfers of sediment.

Much of the landscape in Figure **3** is the result of sea level change triggered by long-term changes in the water cycle.

volcanic basalt   coastal plain   'new' vegetation   isolated stack

 Figure 3 The volcanic basalt was transported to the coast by rivers, glaciers and the wind, and illustrates the links with other physical systems

Sixty second summary

- The coastal system is an open system, with inputs and outputs, processes and stores, negative feedbacks and dynamic equilibrium.
- The sediment cell is a clear example of the application of systems concepts to the coast.
- Coastal systems do not operate in isolation and have clear links with other systems.

Over to you

Summarise the information in Figure **1**. Write **each** definition and example on separate cards and then match **each** of them to a systems term to create **eight** groups.

You need to know:

- the sources of energy and factors affecting waves
- the formation and types of waves
- the role of the tides and currents
- the concept of high and low energy coastlines and the role of wave refraction.

Student Book
pages 106–9

The Sun and wind – the energy behind the waves

The primary source of energy for all natural systems is the Sun. At the coast, waves are the main form of energy. Although waves can be generated by tectonic activity or underwater landslides creating tsunamis, they are mostly formed by the wind.

A number of factors affect wave energy:

- the strength of the wind
- the duration of the wind
- the fetch. The longer the fetch, the more powerful the waves. Figure **1** shows that the UK's longest fetch is over 3000 km to Brazil. This is the same direction as the prevailing wind, so the south and west-facing coasts are often affected by high-energy waves.

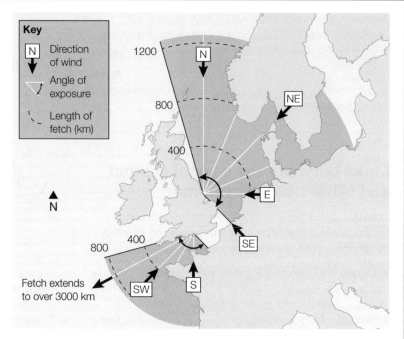

Key
N — Direction of wind
— Angle of exposure
— Length of fetch (km)

Fetch extends to over 3000 km

▲ **Figure 1** *This shows the long fetch to the SW and the N of the UK, accounting for high energy waves*

How are waves formed?

As air moves across the water, friction forms ripples or waves. In the open sea, there is orbital circular motion of the water particles, with little horizontal movement. As waves break on the beach, horizontal movement of water does occur.

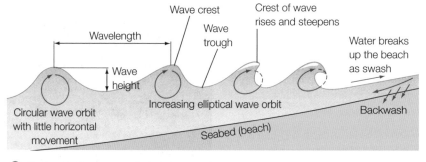

Wavelength
Wave crest
Wave trough
Wave height
Crest of wave rises and steepens
Water breaks up the beach as swash
Circular wave orbit with little horizontal movement
Increasing elliptical wave orbit
Backwash
Seabed (beach)

▲ **Figure 2** *How water particles are affected as waves approach the shore*

Different types of wave

There are two main types of wave: **constructive** and **destructive**.

Characteristic	Constructive waves	Destructive waves
Formation	Distant weather systems generate these waves in the open ocean	Local storms are responsible for these waves
Wave form	Low, surging waves – with a long wavelength	High, plunging waves – with a short wavelength
Wave break	Strong swash, weak backwash	Weak swash, strong backwash
Beach gain/loss	Beach gain (constructive)	Beach loss (destructive)
Beach profile	Usually associated with a gentle beach profile – although, over time, they will build up the beach and make it steeper	Usually associated with a steeper beach profile – although, over time, they will flatten the beach

▲ **Figure 3** *Take time to learn about the two main wave types and their characteristics*

High- and low-energy coasts

Coastline	Characteristics	Where in UK?	Typical landforms
High-energy	Rocky; erosion exceeds deposition	Atlantic-facing coasts, e.g. Cornwall, NW Scotland	Headlands, cliffs, wave-cut platforms
Low-energy	Sandy; deposition exceeds erosion	Sheltered bays and estuaries, e.g. Lincolnshire	Beaches, spits, coastal plains

⬣ **Figure 4** *Types and characteristics of coastline*

Wave refraction

Wave refraction is the distortion of wave fronts as they approach an indented shoreline. This causes energy to be concentrated at headlands and dissipated in bays, which explains why erosive features are found at headlands (cliffs, stacks) and deposition features in bays (beaches). Negative feedback now operates:

* variations in rock strength helped form headlands and bays
* refraction now encourages erosion of the headlands and deposition in the bays.

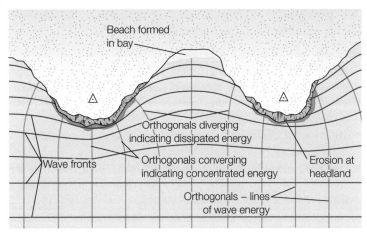

⬣ **Figure 5** *Wave refraction; notice how the shape of the coast distorts the waves affecting energy at the coast*

Tides and currents

Tides are changes in the water level of seas and oceans caused by the gravitational pull (a source of energy) of the moon and, to a lesser extent, the Sun.

* *Tidal range* is the difference in height between high and low tides.
* *Spring* and *neap* tides affect tidal range.
* A high tidal range creates powerful tidal currents (important sources of energy) as tides rise and fall.
* Tidal currents are important transfer mechanisms in transporting sediment.

Rip currents

Rip currents form when plunging waves cause a build-up of water at the top of the beach. The backwash is forced just below the surface of the breaking waves, which can drag people into deep water.

Beaches and waves: negative feedback

Constructive waves are usually associated with relatively gentle beach profiles, enabling waves to surge a long way up the beach. However:

* the profile steepens as more beach material is deposited
* waves become more destructive (plunging rather than surging),
* material removed from the beach is deposited offshore
* the profile becomes less steep, encouraging constructive waves to form.

This 'toing and froing' will result in a state of *dynamic equilibrium*.

⏱ **Sixty second summary**

* Wave energy is affected by several factors including the strength and duration of the wind, and fetch.
* Breaking waves interrupt the normal orbital motion of water particles in open water.
* Constructive waves drive sediment onto the shore; destructive waves erode beach material.
* Tides and currents affect energy concentration and sediment transfer.
* High- and low-energy coastlines are usually associated with erosional and depositional landforms respectively.
* Wave refraction concentrates energy at headlands and dissipates it in bays – demonstrating negative feedback.

 Over to you

Draw a spider diagram to summarise the different aspects of wave energy at the coast.

You need to know:

- the sources of sediment
- about the concept of the sediment cell and its application to the UK
- about sediment budgets and links to physical landforms and human activity.

Student Book pages 110–13

Sources of sediment

The main sediment sources are as follows:

- *Rivers* – the main source, especially in high-rainfall environments where active river erosion occurs.
- *Cliff erosion* – important in areas of relatively soft or unconsolidated rocks.
- *Longshore drift* – transportation from one stretch of coastline (as an output) to another (as an input).
- *Wind* – wind-blown sand can be deposited as sand dunes; they are both accumulations (sinks) of sand and potential sources.
- *Glaciers* – ice shelves *calve* into the sea, depositing sediment within the ice.
- *Offshore* – offshore sediment can be transferred into the coastal zone by waves, tides, currents and storm surges.

Big idea

The transfer of sediment is a major element of the coastal system, influencing human activity and the formation of physical landforms.

Sediment cells

A *sediment cell* is a stretch of coastline, usually between two headlands, where the movement of sediment is more or less contained.

Figure **1** shows a sediment cell system with:

- inputs (sources) from river and coastal erosion, and offshore (e.g. deposits such as bars)
- transfers (flows) such as longshore drift and onshore/offshore processes (e.g. rip currents)
- stores (sinks) such as beaches, sand dunes and offshore deposits (e.g. banks and bars).

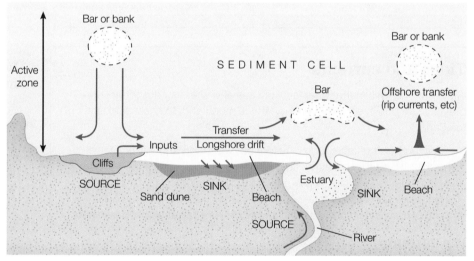

▲ *Figure 1 Inputs (sources), transfers (flows) and stores (sinks) in a sediment cell*

Sediment cells in England and Wales

Sediment cells can be divided into smaller subcells. One of these subcells lies in Christchurch Bay in Dorset (Figure **2**).

It is possible to identify:

- sediment sources (inputs), e.g. cliffs to the west of Barton
- areas of deposition (stores or sinks), e.g. The Shingles
- transfer mechanisms, such as longshore drift operating from west to east.

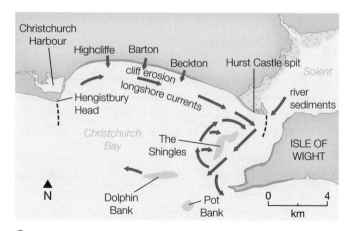

▲ *Figure 2 Notice how sediment is largely confined to this sediment cell in Christchurch Bay*

Sediment budgets

Material in a sediment cell can be considered in the form of a *sediment budget*, with losses and gains. Losses from the system involve deposition in sediment sinks, whereas gains tend to involve coastal erosion or sediment brought into the system by rivers or from offshore sources.

Figure **3** shows the main inputs from coastal erosion, together with the transfers by longshore drift along the coast in East Anglia. By comparing the values of sediment movement, it is possible to see where losses and gains are made at points around the coast.

> 📖 Read a study of the sediment budget in South Carolina on page 112 of the student book.

> **Key**
> All values in thousand m³/yr
> 65 → Direction and amount of sediment moved
> ➡ Input of sediment
> ⬅ Removal of sediment

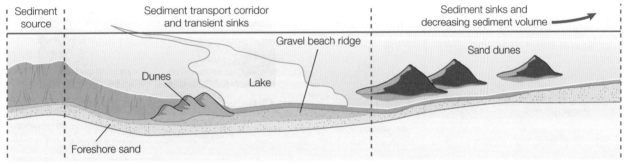

Figure 3 *Coastal sediment budget for East Anglia; notice the quantities of transported sediment around the coast*

The impact of coastal protection

Figure **4** shows a stretch of coastline before and after coastal protection. Coastal protection measures can significantly disrupt the sediment cell and affect the sediment budget.

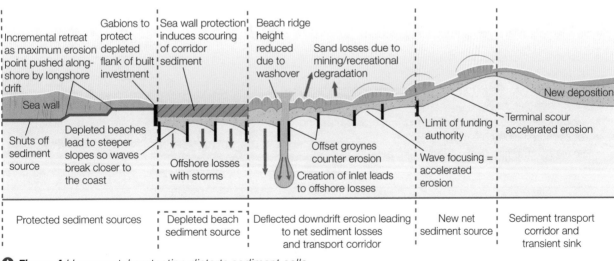

 **Figure 4** *How coastal protection distorts sediment cells*

⏱ **Sixty second summary**

- Rivers are the main sources of coastal sediment. Other sources include cliff erosion and offshore marine erosion.
- Sediment cells are largely self-contained natural coastal systems.
- Distinct inputs, processes and outputs can be identified in sediments cells.
- Sediment budgets involve the quantification of components of the sediment cell providing data, for example, for coastal management.

✏ **Over to you**

Identify the main sources of sediment at the coast and briefly explain the role of **each** one.

Student Book
pages 114–17

You need to know:

- the characteristics and importance of weathering processes at the coast
- the characteristic processes and effects of mass movement
- the role of runoff as a link between the water cycle and the coastal system.

Weathering

Weathering is the breakdown or disintegration of rock in its original place, at or close to the ground surface. As a process, weathering leads to the transfer (flow) of material. There are also important links with other natural systems, such as the water cycle (e.g. freeze–thaw) and the carbon cycle (e.g. carbonation). Weathering can be divided into mechanical, biological and chemical.

Mechanical (physical) weathering

The break-up of rocks without any chemical changes taking place.

- *Frost shattering* (also known as *freeze–thaw*): occurs when water freezes in a crack or joint in the rock, and expands – widening the crack and breaking off rocks which fall to the base of the cliff as scree.
- *Salt crystallisation*: when salt water evaporates, it leaves salt crystals behind. These can exert stresses in the rock, just as ice does, causing it to break up.
- *Wetting and drying*: some rocks expand when wet and contract as they dry. This can cause them to crack and break up.

⚫ **Figure 1** *A major rockfall at the White Cliffs of Dover in 2012, caused by frost shattering*

Chemical weathering

A chemical reaction where salts may be dissolved or result in a clay-like deposit which is then easily eroded.

- *Carbonation* – rainwater absorbs carbon dioxide from the air to form weak carbonic acid, which reacts with calcium carbonate in rocks to form calcium bicarbonate, which is easily dissolved.
- *Oxidation* – the reaction of minerals with oxygen leaving rocks more vulnerable to weathering.
- *Solution* – the dissolving of rock minerals.

Biological weathering

The breakdown of rocks by organic activity.

- Thin plant roots grow into small cracks. As the roots grow, they expand and exert force on the cracks, breaking up the rock.
- Birds and animals dig burrows into cliffs.
- Marine organisms are also capable of burrowing into rocks or of secreting acids.

Mass movement

The downhill movement of material under the influence of gravity. It is common at the coast – the sheer weight of rainwater, combined with weak geology, is the major cause of cliff collapse.

Mass movement can be classified into *creep*, *flow*, *slide* and *fall*. Each process represents a flow or transfer of material and is an output from one store (land) and an input to another store (beach/sea).

Type of mass movement	Nature of movement	Rate of movement	Wet/ dry
Soil creep Solifluction	Creep/flow	Imperceptible	Wet
Mudflow	Flow	Often quite rapid	Wet
Runoff	Flow	Rapid	Wet
Landslide/debris slide Slump/slip	Slide	Usually rapid	Dry Wet
Rockfall	Fall	Rapid	Dry

⚫ **Figure 2** *Classifying mass movement*

Soil creep

Particles rise towards the ground surface due to wetting or freezing. They then return vertically to the surface in response to gravity as the soil dries out or thaws.

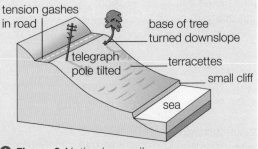

▲ *Figure 3 Notice how soil creep distorts trees and telegraph poles*

Mudflows

The flowing downhill of earth and mud, often after heavy rainfall. Water gets trapped within the rock, forcing rock particles apart leading to slope failure.

Figure 4 ◗
Mudflows form distinctive lobes

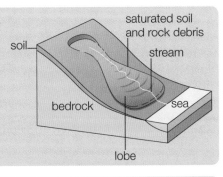

Rockfall

The sudden breaking away of individual rocks at a cliff face, associated with steep cliffs and heavily jointed, resistant rock. They are often triggered by mechanical weathering or an earthquake. Resulting scree (talus) is gradually removed by the sea – an input into the sediment cell.

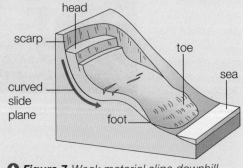

▲ *Figure 6 Isolated rocks fall off a cliff face*

Landslip or slump

Similar to a landslide except its slide surface is curved rather than flat. Landslips commonly occur where permeable rock overlies impermeable rock, which causes a build-up of pore water pressure.

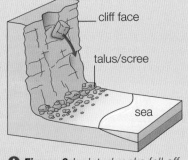

▲ *Figure 7 Weak material slips downhill*

Landslide

A block of rock moving rapidly downhill along a slide plane (often a bedding plane roughly parallel to the surface). The moving block of material remains largely intact (unlike a mudflow). Landslides are triggered by earthquakes or heavy rainfall, when the slip surface becomes lubricated and friction is reduced.

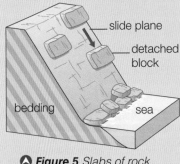

▲ *Figure 5 Slabs of rock slide down a cliff face*

Solifluction

Solifluction is similar to soil creep but specific to cold periglacial environments. Soil thaws in summer and becomes saturated because it overlays impermeable permafrost. This **active layer** slowly moves downhill by a combination of heave and flow.

Runoff

Runoff links the water cycle and the coastal system. Overland flow moves particles into the coastal zone, potentially forming an input into the sediment cell. Runoff transfers both water and sediment from one store (the rock face) to another (a beach/the sea).

 Sixty second summary

- Weathering is the breakdown of rocks at or close to the ground.
- Mechanical weathering creates angular sediment; chemical weathering creates fine clay. Biological weathering can create either.
- Mass movement can result in huge quantities of sediment and rock being released into the coastal zone to be eroded and transported by the sea.
- Mass movement processes include landslides, rockfalls and slumps.
- Runoff is a link between the water cycle and the coastal system.

Over to you

Practise sketching and annotating simple diagrams to identify the main characteristics of the processes of mass movement.

Student Book
pages 118–21

You need to know:

- processes of erosion and factors affecting the nature and rate of erosion
- processes of transportation and the mechanisms and role of longshore drift
- conditions leading to coastal deposition.

Coastal erosion

Coastal erosion plays a vital role in the coastal system, removing debris from the foot of cliffs and providing an input into coastal sediment cells.

There are several distinct processes of coastal erosion. These processes rarely operate in isolation and will work together to erode a stretch of coastline. Figure **1** shows the result of some of these processes.

Hydraulic action

This is the compression and subsequent expansion of air forced into cracks in rock by wave pounding. This continuous process causes rock to break off.

Cavitation is the implosion of tiny bubbles in the water. This generates tiny jets of water which causes erosion.

Factors affecting coastal erosion

There are several factors that affect the nature and rate of coastal erosion:

- *Waves* – most erosion happens during winter storms when destructive waves are at their most powerful.
- *Rock type (lithology)* – tough and resistant rocks (e.g. granite) erode at very slow rates compared to weaker clays and shales.
- *Geological structure* – cracks, joints, bedding planes and faults create weaknesses in a cliff that can be exploited by erosion (Figure **1**).
- *Beach* – beaches absorb wave energy and reduce the impact of waves on a cliff.
- *Subaerial processes* – weathering and mass movement create debris that is easily eroded by the sea.
- *Coastal management* – groynes and sea walls impact on sediment transfer and patterns of wave energy (see 3.9).

Wave quarrying

The scooping action of waves breaking against unconsolidated material such as sands and gravels.

Corrasion

Sand and pebbles are hurled by wave action at the cliff foot, chipping away at the rock.

Abrasion

A 'sandpapering effect' of waves dragging sediment up and down or across the shoreline, eroding and smoothing rocky surfaces.

Solution (corrosion)

Weak acids in seawater dissolve alkaline rock (e.g. chalk or limestone), or the alkaline cement that bonds rock particles together.

⬥ **Figure 1** *Cliff weaknesses are exploited by coastal erosion, Flamborough Head, Yorkshire*

Coastal transportation

It is possible to identify four methods of transportation. The key factors affecting the type of transportation are velocity (energy) and particle size (mass):

- *Traction* – the rolling of coarse sediment along the sea bed.
- *Saltation* – sediment is 'bounced' along the seabed.
- *Suspension* – smaller (lighter) sediment picked up and carried within the flow of the water.
- *Solution* (dissolved load) – chemicals dissolved in the water and transported elsewhere.

Longshore (littoral) drift

Longshore drift is an important transfer (flow) mechanism which moves vast amounts of sediment:

① Most waves approach a beach at an angle, carrying material up the beach indirectly (obliquely).
② The backwash then pulls material down the beach at right angles to the shore (due to gravity – a source of energy).
③ Management strategies can interrupt these transfers, depriving beaches of material and increasing erosion.

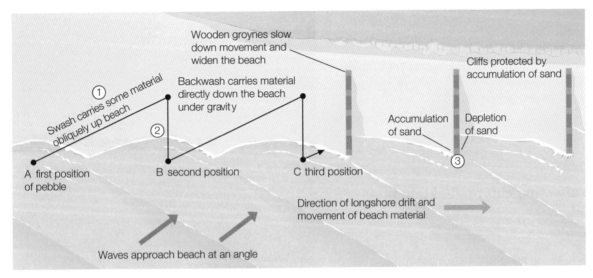

Wooden groynes slow down movement and widen the beach

Cliffs protected by accumulation of sand

① Swash carries some material obliquely up beach

Backwash carries material directly down the beach under gravity

②

Accumulation of sand   Depletion of sand

③

A first position of pebble

B second position

C third position

Direction of longshore drift and movement of beach material

Waves approach beach at an angle

⬆ **Figure 2** *The effect of groynes on longshore drift*

Coastal deposition

Deposition takes place when the velocity of the water (or wind) falls below a critical value for a particular size of particle:

- In high-energy environments, clay and sand will be easily transported away leaving behind the larger, coarser pebbles that form shingle beaches. For example, the exposed parts of the south coast of England.
- In low-energy environments, such as river estuaries, the very smallest clay particles will eventually settle to form mudflats.

⬆ **Figure 3** *A sandbank at Ko Pada Island, Thailand – an example of a sediment store or 'sink'*

 Sixty second summary

- Erosion is crucial to the coastal system. It removes debris and provides inputs to sediment cells.
- Hydraulic action, quarrying and abrasion are important erosional processes.
- Factors of erosion include wave energy, rock type, geological structure and human management.
- Processes of transportation include traction, saltation, suspension and solution.
- Longshore drift is an important process in moving sediment along a coastline. Its disruption can exacerbate erosion.
- Deposition occurs when velocity falls in sheltered locations such as bays and estuaries.

 Over to you

Summarise the key marine processes of erosion, transportation and deposition.

Student Book
pages 122–5

You need to know:
- the characteristic features and formation of landforms and landscapes of coastal erosion
- the factors and processes that affect their development.

Landforms of coastal erosion

Cliffs and wave-cut platforms

When waves break against the foot of a *cliff*, erosion (hydraulic action and corrasion in particular) tends to be concentrated close to the high-tide line. This creates a *wave-cut notch* (Figures **1** and **2**). As the notch gets bigger, the cliff is undercut and the rock above it becomes unstable, eventually collapsing.

As these erosional processes are repeated, the notch migrates inland and the cliff retreats (Figure **2**), leaving behind a gently sloping *wave-cut platform.*

Factors affecting cliff profiles and rate of retreat

Steep cliffs tend to occur where:

- the rock is strong and resistant to erosion
- sedimentary rocks dip steeply or even vertically
- there is no beach and the coastline is exposed
- the coastline is exposed to a long fetch and high-energy waves.

Gentle cliffs tend to occur where:

- rocks are weak or unconsolidated and prone to slumping
- rocks dip towards the sea
- location is sheltered with a short fetch and low-energy waves
- a wide beach absorbs wave energy, preventing erosion.

Other factors leading to rapid rates of cliff retreat include rising sea levels and human activities (such as coastal defences elsewhere leading to increased erosion).

Note the staining which identifies submersion within the full tidal range (up to the highest spring tides)

⬆ **Figure 1** *A wave-cut notch at Flamborough Head in Yorkshire*

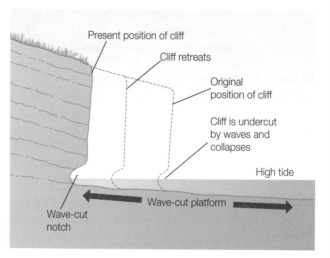

⬆ **Figure 2** *Cliff retreat. Notice that a wave-cut platform will only be exposed at low tide.*

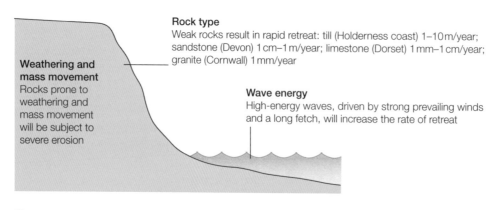

Rock type
Weak rocks result in rapid retreat: till (Holderness coast) 1–10 m/year; sandstone (Devon) 1 cm–1 m/year; limestone (Dorset) 1 mm–1 cm/year; granite (Cornwall) 1 mm/year

Weathering and mass movement
Rocks prone to weathering and mass movement will be subject to severe erosion

Wave energy
High-energy waves, driven by strong prevailing winds and a long fetch, will increase the rate of retreat

⬆ **Figure 3** *Several factors affect the rate of retreat*

The sequence of diagrams in Figure **4** shows that cliff profiles often reflect geological structure such as the dip of rock strata.

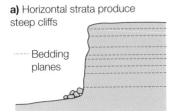

a) Horizontal strata produce steep cliffs

----- Bedding planes

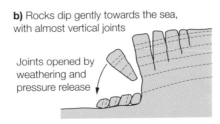

b) Rocks dip gently towards the sea, with almost vertical joints

Joints opened by weathering and pressure release

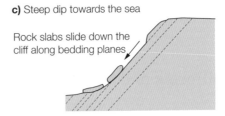

c) Steep dip towards the sea

Rock slabs slide down the cliff along bedding planes

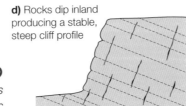

d) Rocks dip inland producing a stable, steep cliff profile

e) Rocks dip inland but with well-developed joints at right angles to bedding planes

Joints act as slide planes

Figure 4 ◗

Notice how geological structure affects cliff profiles

Caves, arches, stacks and stumps

Caves, arches, stacks and stumps are all part of a sequence of coastal landform development:

- Joints and faults are eroded by hydraulic action and abrasion, to form *caves*.
- If the overlying rock then collapses, a *blowhole* will develop.
- If two caves join, or a cave is eroded through a headland, an *arch* is formed.
- The base of the arch is widened by erosion and weathering.
- Eventually, the arch collapses leaving an isolated pillar of rock called a *stack*.
- The stack eventually collapses to leave a *stump*.

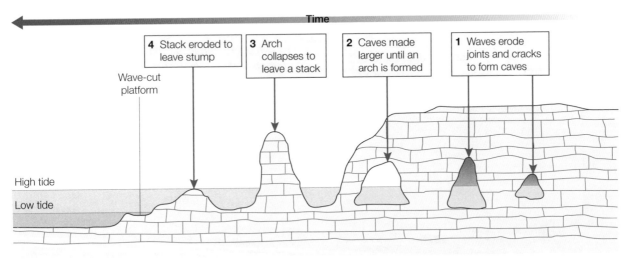

Time

4 Stack eroded to leave stump

3 Arch collapses to leave a stack

2 Caves made larger until an arch is formed

1 Waves erode joints and cracks to form caves

Wave-cut platform

High tide

Low tide

🔺 **Figure 5** *The sequence of formation of caves, arches, stacks and stumps*

Sixty second summary

- Landforms are individual components of the coast, but should be considered in the context of the broader coastal landscape.
- Processes of erosion (particularly hydraulic action) – cause cliffs to become eroded. As they retreat they leave behind a wave-cut platform.
- Several factors affect cliff profiles and the rate of cliff retreat, such as wave energy, rock type and geological structure.
- Headlands are eroded to form a sequence of distinctive coastal landforms – caves, arches, stacks and stumps.

Over to you

Use simple annotated diagrams to describe the formation of landforms of coastal erosion studied in this section.

You need to know:

- what role coastal deposition has in the formation of coastal landforms
- the key characteristics of landforms of coastal deposition
- about the development of vegetation successions on sand dunes and mudflats.

Student Book
pages 126–33

Landforms of coastal deposition

Deposition occurs along the coast when waves no longer have enough energy to transport sediment.

Beaches

A beach is an important temporary store in the coastal system, and extends from approximately the highest high tide to the lowest low tide.

- Beach **accretion** will take place during a prolonged period of constructive waves driven by storms many hundreds of miles away.
- Destructive waves, resulting from localised storms, may remove vast quantities of sediment, even exposing previously covered wave-cut platforms.

Swash-aligned and drift-aligned beaches

Beaches can be described as being **swash-aligned** or **drift-aligned** depending on their orientation relative to the prevailing wind (and wave) direction (see Figure **1**).

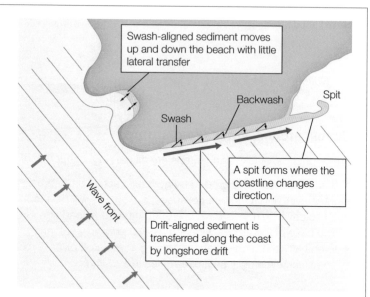

Swash-aligned sediment moves up and down the beach with little lateral transfer

A spit forms where the coastline changes direction.

Drift-aligned sediment is transferred along the coast by longshore drift

🔺 **Figure 1**

Beach characteristics (drift-aligned or swash-aligned) are determined by coastal orientation and the angle of approaching waves

Characteristics of beaches

Beaches often form part of a much broader area of deposition extending into the offshore zone. (Figure **2**).

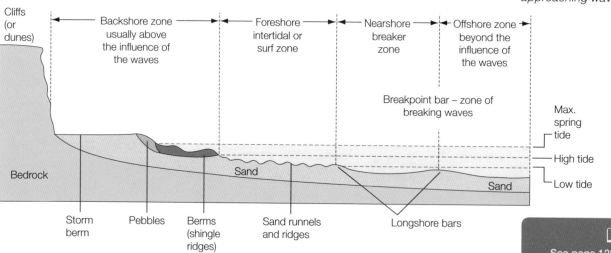

🔺 **Figure 2** *Zones and typical features of a beach. There are often several **berms** representing different tidal levels. There may also be a storm berm at the highest point on a beach.*

📖
See page 128 of the student book for the factors affecting beach profiles.

Spits

A *spit* is a long, narrow feature, made of sand or shingle, that extends from the land into the sea (or part of the way across an estuary). Spits form on drift-aligned beaches (see Figure 1).

Tombolos

A **tombolo** is a beach (or ridge of sand and shingle) between a small island and the mainland. Chesil Beach is an example of a tombolo, linking the Isle of Portland with Weymouth on the mainland.

Offshore bars

Also known as sandbars, **offshore bars** are submerged or partly exposed ridges of sediment created by offshore waves. Offshore bars act as both sediment sinks and, potentially, sediment input stores.

Figure 3 ◗

Destructive waves are responsible for the formation of offshore bars

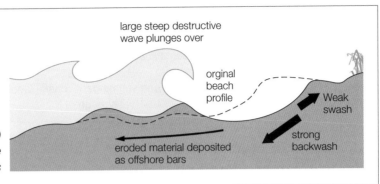

large steep destructive wave plunges over

orginal beach profile

Weak swash

strong backwash

eroded material deposited as offshore bars

Sand dunes

Sandy beaches may be backed by *sand dunes*, formation of which needs:

- large quantities of sand from constructive waves
- large tidal range, creating a large exposure of sand that can dry out at low tide
- dominant onshore winds, to blow dried sand to the back of the beach.

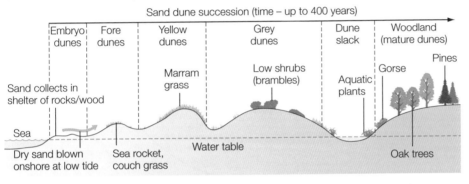

Sand dune succession (time – up to 400 years)

Embryo dunes | Fore dunes | Yellow dunes | Grey dunes | Dune slack | Woodland (mature dunes)

Pines

Marram grass

Low shrubs (brambles)

Gorse

Aquatic plants

Sand collects in shelter of rocks/wood

Sea

Water table

Dry sand blown onshore at low tide

Sea rocket, couch grass

Oak trees

⬆ **Figure 4** *The stages in the development of sand dunes and plant succession. Dunes darken in colour as decaying plants add organic matter.*

Barrier beaches (bars)

Where a beach or spit extends across a bay to join two headlands, it forms a barrier beach or *bar*, e.g. the barrier beach at Start Bay in Devon.

Barrier beaches and bars can also trap water behind them to form lagoons, such as Slapton Ley in Devon.

Estuarine mudflats and saltmarshes

Due to the low velocities in estuaries, expansive *mudflats* can form over time. They then develop into **saltmarshes** – areas of flat, silty sediments that accumulate around estuaries or lagoons. They develop in three types of environment:

- sheltered areas where deposition occurs (e.g. in the lee of a spit)
- where salt and freshwater meet (e.g. estuaries)
- where there are no strong tides or currents to prevent sediment deposition and accumulation.

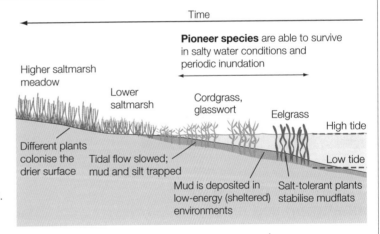

Time

Pioneer species are able to survive in salty water conditions and periodic inundation

Higher saltmarsh meadow

Lower saltmarsh

Cordgrass, glasswort

Eelgrass

High tide

Low tide

Different plants colonise the drier surface

Tidal flow slowed; mud and silt trapped

Mud is deposited in low-energy (sheltered) environments

Salt-tolerant plants stabilise mudflats

⬆ **Figure 5** *This describes the vegetation succession on a saltmarsh*

- Deposition is responsible for the formation of coastal landforms such as beaches, spits and bars.
- Beaches display characteristic features including **berms**, **cusps** and **runnels**.
- Deposition by the sea combines with transportation and deposition by wind to form sand dunes.
- Deposition of mud in estuaries forms mudflats and saltmarshes.
- Vegetation develops through processes of succession on sand dunes and mudflats.

Summarise in a table the key characteristics and formative processes associated with the landforms of coastal deposition.

You need to know:

- the causes and consequences of eustatic and isostatic change
- the role of tectonics in sea level change
- the characteristics of emergent coastal and submergent coastal landforms
- the potential impacts of contemporary sea level change.

Student Book
pages 134–9

Long-term sea level change

Sea levels have been very different in the past compared with the present day. During the Quaternary period, sea levels fell during glacial periods, and rose during the warm (interglacial) periods, due to the nature of precipitation.

Eustatic and isostatic change

There are two main causes of sea level change:

- **eustatic change** (sea level itself rises or falls). In glacial periods, water that is usually held in the oceans is stored in ice sheets. As a result, sea levels fall. As the ice sheets melt, their stored water flows into rivers and the sea, and sea levels rise.
- **isostatic change** (land itself rises or falls, relative to the sea). Isostatic change occurs locally. During glacial periods, the enormous weight of the ice sheets makes the land sink. As the ice melts, the reduced weight of the ice causes the land to readjust and rise (*isostatic recovery*).

The role of tectonic activity

Past tectonic activity has had a direct impact on sea levels, due to:

- the uplift at destructive and collision plate margins resulting in a relative fall in sea level in some parts of the world
- local tilting of land at destructive margins, e.g. some ancient Mediterranean ports have been submerged, whilst others have been stranded above the current sea level.

 Read about the impacts of the 2004 Boxing Day earthquake on sea levels on page 136 of the student book.

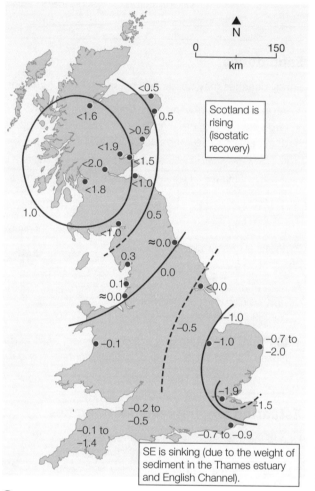

Figure 1 *Isostatic change in the UK. Parts of the UK are rising (positive mm/year) or falling (negative mm/year).*

Landforms caused by changing sea level

A fall in sea level exposes land previously covered by the sea, creating an *emergent coastline*. A rise in sea level floods the coast and creates a *submergent coastline*.

Emergent coastal landforms

Raised beaches are common on the west coast of Scotland, backed by eroded cliff lines (*relic cliffs*), with wave-cut notches and caves as evidence of past marine erosion. Wave-cut (marine) platforms can also be exposed if sea levels fall sufficiently.

Figure 2 *Relic cliffs, raised beaches and raised wave-cut platforms are landforms associated with falling sea levels*

Submergent coastal landforms

Rias are sheltered, winding inlets formed when a rise in sea level drowned river valleys. They are common in south-west England where sea levels rose after the last ice age, e.g. the Kingsbridge Estuary in Devon.

Fjords are formed when deep glacial troughs are flooded by a rise in sea level. They are long and steep-sided, with a U-shaped cross-section. Unlike rias, fjords are much deeper inland than they are at the coast.

Dalmatian coasts form when valleys running parallel to the coast are flooded by a rise in sea level. The exposed tops of the ridges form a series of offshore islands running parallel to the coast.

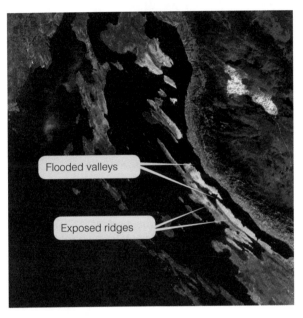

Flooded valleys

Exposed ridges

Figure 3 🔾
A satellite image of the Dalmatian coast in Croatia

Contemporary sea level change

Sea levels have recently started changing:

- According to the Intergovernmental Panel for Climate Change (IPCC), sea levels stabilised about 3000 years ago and have changed little since then until very recently.
- From the late nineteenth century to the late twentieth century, sea levels rose globally by about 1.7 mm per year. This has increased to about 3.2 mm per year in the period 1993–2010.
- The IPCC estimates that by 2100 sea levels could rise by between 30 cm and 1 m. Sea level rise is mainly the result of thermal expansion of water, due to warming, and the melting of *freshwater* ice, such as ice sheets and mountain glaciers. Between 1880 and 2010, global temperatures rose by an average of 0.85 °C.

The impact of climate change in Kiribati

The nation of Kiribati in the Pacific Ocean consists of 33 islands of low-lying sand and mangrove atolls. It has been predicted that many of its islands could disappear under the sea in the next 50 years.

What next for Kiribati?

Rising sea levels in Kiribati are contaminating its groundwater sources and therefore agriculture. So in 2014, Kiribati purchased land on one of the islands of Fiji – 2000 km from Kiribati. This land will be used in the immediate future for agriculture and fish-farming projects – to guarantee the nation's food security. In the longer term, people could move there.

If the islands are submerged, Kiribati's population will become *environmental refugees* – people forced to migrate as a result of changes to the environment.

Sixty second summary

- Sea levels respond to climate change, in particular to the onset of glacial and interglacial periods.
- Eustatic change is changes in sea level; isostatic change is rise and fall of the land relative to the sea.
- Tectonic activity can impact sea level.
- Falls in sea level create emergent landforms, such as raised beaches and relic cliffs.
- Rising sea levels create submergent landforms, such as rias and fjords.
- Recent sea level rise associated with climate change is threatening many low-lying islands and coastal regions.

Over to you

Make sure you know the difference between eustatic and isostatic. Group the landforms on this page into those associated with emergent and submergent coastlines.

Student Book
pages 140–3

You need to know about:

- the forms of traditional coastal management – hard- and soft-engineering
- sustainable integrated approaches such as Shoreline Management Plans (SMP) and Integrated Coastal Zone Management (ICZM).

Human intervention at the coast

According to the United Nations Environment Programme (UNEP), about half the world's population live within 60 km of a coast and three-quarters of all large cities are at the coast.

- The demand for increasing development puts coastal areas under immense pressure. This results in environmental damage, habitat destruction (e.g. mangrove forests, sand dune systems, coral reefs) and pollution of coastal waters.
- Large numbers of people living at the coast are at risk of flooding and coastal erosion. This risk is likely to increase as global temperatures increase and sea levels rise.

 Big idea

Coastal management involves how to balance the potentially harmful effects of natural processes with the increasing developmental demands made by people.

Traditional approaches

Hard-engineering approach

Type of structure	Description	Advantages	Disadvantages
Groynes	Timber or rock structures (built at right angles to the coast) trap sediment being moved along the coast by longshore drift.	• Work with natural processes to build up the beach. • Increases tourist potential and protects the land behind it. • Not too expensive.	• Starve beaches further along the coast of fresh sediment, leading to increased erosion elsewhere. • Unnatural and can be unattractive.
Sea walls	Stone or concrete walls at the foot of a cliff, or at the top of a beach. They usually have a curved face to reflect waves back into the sea.	• Effective prevention of erosion. • They often have a promenade for people to walk along.	• Reflect wave energy, rather than absorbing it. • Intrusive and unnatural looking. • Very expensive to build and maintain.
Rip rap (rock armour)	Large rocks placed at the foot of a cliff, or at the top of a beach. It forms a permeable barrier to the sea – breaking up waves, but allowing some water to pass through.	• Relatively cheap and easy to construct and maintain.	• Very intrusive and unsympathetic to local geology. • Can be dangerous for people clambering over them.
Revetments	Sloping wooden, concrete or rock structures placed at the foot of a cliff or the top of a beach. They break up the waves' energy.	• Relatively inexpensive to build (up to £4500/m).	• Intrusive and very unnatural looking. • Need high levels of maintenance.
Offshore breakwater	A partly submerged rock barrier, designed to break up the waves before they reach the coast.	• An effective permeable barrier.	• Visually unappealing. • Potential navigation hazard.

⬆ *Figure 1 This table shows how each hard-engineering structure helps to protect the coast, as well as the advantages and disadvantages*

Soft-engineering approach

Method	Description	Advantages	Disadvantages
Beach nourishment	The addition of sand or pebbles to an existing beach. The sediment is usually dredged from the nearby seabed.	• Relatively cheap and easy to maintain. • Looks natural and increases tourist potential by creating a bigger beach.	• Needs constant maintenance because of the natural processes of erosion and longshore drift.
Cliff regrading and drainage	Cliff regrading reduces the angle of the cliff to help stabilise it. Drainage removes water to prevent landslides and slumping.	• Most effective on clay or loose rock. • Drainage is cost-effective.	• Regrading effectively causes the cliff to retreat. • Drained cliffs can dry out and lead to collapse (rockfalls).
Dune stabilisation	Marram grass can be planted to stabilise dunes. Areas can be fenced to keep people off newly planted dunes.	• Maintains a natural coastal environment with important wildlife habitats. • Relatively cheap and sustainable.	• Time consuming to plant. • People may respond negatively to being kept off certain areas.
Marsh creation	A form of managed retreat allowing low-lying coastal areas to be flooded by the sea to become saltmarsh.	• Relatively cheap. • Creates a natural buffer to powerful waves and an important wildlife habitat.	• Agricultural land is lost. • Farmers or landowners need to be compensated.

Figure 2 *This table shows how each soft-engineering approach helps to protect the coast, as well as the advantages and disadvantages*

Sustainable integrated approaches

Soft-engineering schemes are intended to work with nature over a long period of time, but they tend to focus on a particular stretch of coastline. Increasingly, management authorities are looking at holistic (integrated/interconnected) management plans for significant stretches of coastline.

Shoreline Management Plans (SMP)

An SMP has been written for each of the sediment cells in England and Wales (see 3.1). They are based on the principle that intervention will have few knock-on effects outside each cell. Four options are considered for any stretch of coastline:

- *Hold the line* – maintaining the current coastline
- *Advance the line* – extending the coastline out to sea
- *Managed retreat/strategic realignment* – allowing the coastline to retreat in a managed way
- *Do nothing/no active intervention* – letting nature take its course.

Integrated Coastal Zone Management (ICZM)

ICZM brings together all of the stakeholders involved in the development, management and use of the coast in order to manage the coastal environment based on natural process and the coastal system.

Adopting an ICZM strategy means that complete sections of coast are managed holistically, enabling sustainable measures to be implemented based on natural process and the coastal system.

Read about EU initiatives on coastal management on page 143 in the student book.

Sixty second summary

- Hard-engineering strategies use artificial structures which interfere with natural processes.
- Soft-engineering strategies work with natural processes.
- SMP provide a comprehensive planning overview of natural processes and human development needs.
- ICZM involves the holistic consideration of coastal management issues for a broad coastal zone.

Over to you

Ensure that you have an instinctive understanding of the word 'sustainable' and can explain why soft-engineering approaches are generally more sustainable than hard-engineering methods in protecting coastlines.

Case Study

You need to know:

- the features of the Holderness coastal system
- which factors affect the coastal system (geology, prevailing winds and management)
- the formative processes and landforms associated with the Holderness coastal system.

*Student Book
pages 144–7*

The Holderness coastal system

The Holderness coast is a well-known stretch of coastline in eastern England. It forms a subcell in sediment cell 2 (Figure **2**, 3.1) and essentially comprises three distinct coastal units:

1 Flamborough Head in the north – a chalk promontory with many typical landforms of coastal erosion
2 Bridlington Bay to Spurn Head – a zone of erosion and sediment transfer characterised by a rapid rate of cliff retreat
3 Spurn Head – a classic spit formed at the estuary of the River Humber.

Factors affecting the coastal system

- *Geology* – chalk, a relatively resistant rock, forms a broad arc in the region, stretching from the Lincolnshire Wolds in the south to the coast at Flamborough Head (Figure **1**). Much of the coastal area is glacial till deposited at the end of the last glacial period.
- *Prevailing winds (wave energy)* – powerful waves are driven against the coast when the wind blows from the north-east (direction of greatest fetch). Longshore drift operates from north to south (Figure **1**).
- *Management* – parts of the coastline have been protected with hard-engineering structures. Although these interventions have helped to protect specific localities, such as Hornsea and Mappleton, they have deprived areas further south of sediment, exacerbating coastal erosion.

Big idea

The transfer of sediment is a major element of the coastal system, influencing human activity and the formation of physical landforms.

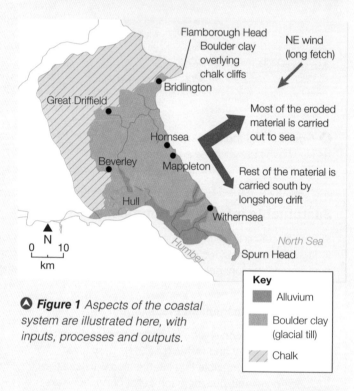

○ **Figure 1** *Aspects of the coastal system are illustrated here, with inputs, processes and outputs.*

Flamborough Head

Flamborough Head is a headland made of chalk – a resistant, sedimentary rock. Joints and faults in the chalk are readily exploited by the processes of weathering and erosion to form narrow clefts in the coastline (Figure **2** and Figure **1**, 3.5).

The sea is actively eroding and undercutting the base of the chalk cliffs leading to frequent rockfalls and forming characteristic cliffs, wave-cut platforms and stacks.

○ **Figure 2** *Landforms of coastal erosion at Flamborough Head*

Bridlington Bay to Spurn Head

With rates of erosion between 1m and 10m per year, the Holderness coast has one of the most rapid rates of erosion in Europe. Several factors account for this:

- long fetch and powerful waves from the north-east
- weak and unconsolidated till cliffs
- extensive mass movement, especially slumping, caused by undercutting and saturation of clay within the cliffs
- narrow beach making the cliffs vulnerable to wave attack and undercutting
- lack of coastal defences.

Although erosion creates threats, a great deal of coarse sediment is transferred by longshore drift to the south, building up beaches and reducing erosion (a negative feedback for the system).

Note the darker till indicating saturation of the clay.

This slump debris will be removed rapidly by wave action and longshore drift, exposing the cliff to further erosion.

▲ **Figure 3** *Cliffs on the Holderness coast. Notice the eroded material at the foot of the cliffs.*

 Case Study

Spurn Head

Spurn Head spit represents a temporary sediment store or sink. Much of the material that forms the spit is derived from the Holderness coast and is transferred south by longshore drift. On reaching the River Humber estuary, the deposited sediment grows out to form a spit.

Spurn Head protects the City of Hull and many villages bordering the River Humber from the effects of storm waves and flooding. It is extremely narrow for much of its length and has frequently been breached and destroyed by major storms.

Following a massive breach in 1849, groynes and revetments (wooden barriers) were erected to stabilise the spit. Since then, some of the defences have fallen into disrepair and the spit remains in serious threat of being breached.

▲ **Figure 4** *The narrowness of the spit at Spurn Head can easily be seen in this photograph*

Sixty second summary

- The Holderness coast demonstrates the typical features associated with a coastal system, such as inputs, processes and outputs.
- Several factors affect the coastal system including geology (presence of resistant chalk and weak glacial till), prevailing winds (from the north-east) and management.
- Flamborough Head (resistant chalk headland) exhibits many features of erosion such as cliffs, wave-cut platforms and stacks.
- Bridlington Bay to Spurn Head is a rapidly eroding coastline.
- Spurn Head is a good example of a spit.

Over to you

Draw a simple, annotated sketch map of the Holderness coast to identify the key aspects of the coastal system and characteristic landforms.

You need to know:

- the location and characteristic features of the Odisha coast
- the economic and environmental pressures on the coast
- the risks associated with the coastline
- management strategies that balance economic demands and environmental concerns.

Student Book
pages 148–53

Odisha: a distinctive and contrasting coastal landscape

Odisha is a state on the eastern coast of India bordering the Bay of Bengal (Figure **1**). The coastline is relatively straight with few natural inlets or harbours. The narrow, level coastal strip known as the Odisha Coastal Plains supports the bulk of the state's population.

- The coast is essentially one of deposition, comprising several major deltas.
- It represents a significant sediment store, providing a source (and sink) of sediment for this part of the Bay of Bengal.
- Rivers provide important transfers of sediment into the region in forming deltaic deposits.
- There is a wide range of coastal and marine flora and fauna (including mangrove forest).
- Chilika Lake is a brackish salty lagoon, renowned for its birdlife. During the monsoon season, the lake is diluted by the freshwater rainfall, and occupies a larger area than during the rest of the year.
- Chilika Lake is a temporary store in the water cycle; the beach that has created the lake is an important store within the coastal system.

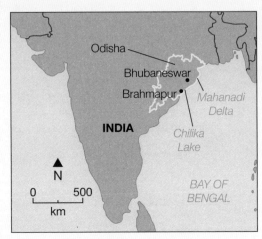

◢ **Figure 1** *Location of Odisha, India*

◢ **Figure 2** *The Odisha coast and the three major coastal ecological environments*

Opportunities for human occupation and development

In addition to providing relatively flat land for settlement, the coastal plain provides several economic and environmental opportunities.

- A wide variety of marine and coastal flora and fauna.
- Large stocks of fish, marine mammals and reptiles.
- Huge potential for offshore wind, tidal and wave power.
- Opportunities for exploiting offshore oil and natural gas, as well as seabed mining.
- Local employment in coastal fishing and aquaculture, such as shrimps.

Case Study

Risks for human occupation and development

In 2011 the Ministry of Environment and Forests released its latest Assessment of Shoreline Change. Rates of erosion have increased in recent decades, partly through natural processes but also as a consequence of human intervention methods.

The Indian government has become concerned about the increased vulnerability of coastal communities to storm surges and tsunamis, as well as the longer-term threats posed by climate change and rising sea levels.

Key findings of the Assessment of Shoreline Change included:

- Around 48% of the coast of Odisha is accreting, with around 38% eroding and 14% stable.
- Most accretion is in the north, focused on the major deltas.
- Most of the erosion is in the south. Here, major hard-engineering structures have interfered with sediment transfer and destabilised patterns of wave energy, increasing localised rates of erosion.

Managing the Odisha coast

A recent ICZM project has promoted the sustainable use of the natural resources of the Odisha coast while maintaining the natural environment.

The ICZM aims to:

- establish sustainable levels of economic and social activity
- resolve environmental, social and economic challenges and conflicts
- protect the coastal environment.

Read more about the ICZM plan for the Odisha coast on page 151 of the student book and management of the Mahanadi Delta on page 152.

The risk from tropical cyclones

The Odisha coast is at risk from tropical cyclones. The frequency and intensity of these cyclones may increase in the future as a result of climate change. Along with a rising sea level, this represents a significant threat to the coast.

Reacting to the 1999 Odisha Cyclone, the authorities now employ a number of mitigation strategies:

- providing relief supplies ahead of an approaching storm
- broadcasting warnings
- conducting staged evacuations away from the most vulnerable areas.

The relatively small death toll (44) of Cyclone Phailin in 2013 shows that these strategies work, as well as the resilience of the people and their ability to adapt to changing circumstances.

▲ *Figure 3* *The aftermath of Cyclone Phailin, 2013*

 Sixty second summary

- Odisha has a dynamic coastal system; it is affected by significant seasonal variation in wave energy and sediment input via the region's major rivers.
- The coast is characterised by a broad lowland plain with extensive features of coastal deposition.
- There are many economic and environmental opportunities, such as mineral exploitation and preservation of coastal ecosystems.
- Erosion due to natural processes and also human intervention is a concern.
- Communities are vulnerable to the threat of tsunamis and storm surges.
- Management strategies range from piecemeal coastal defences to much broader holistic approaches.

 Over to you

Create a table summarising the characteristics, pressures and management of the Odisha coast.

4 Glacial systems and landscapes

Your exam

(AL) *Glacial systems and landscapes* is an **optional topic**. You must answer **one** question in Section B of Paper 1: Physical geography, from a choice of **three**: *Hot desert systems and landscapes* or *Coastal systems and landscapes* or *Glacial systems and landscapes*.
Paper 1 carries 120 marks and makes up 40% of your A Level. Section B carries 36 marks.

(AS) *Glacial systems and landscapes* is an **optional topic**. You must answer **one** question in Section A of Paper 1: Physical geography and people and the environment, from a choice of **three**: *Water and carbon cycles* or *Coastal systems and landscapes* or *Glacial systems and landscapes*.
Paper 1 makes up 50% of your AS Level. Section A carries 40 marks.

Specification subject content (Specification reference in brackets)

Either tick these boxes as a record of your revision, or use them to identify your strengths and weaknesses

Your revision checklist

Section in student book and revision guide	☹	😐	☺	Key terms you need to understand Complete the **key terms** (not just the words in bold) as your revision progresses. 4.1 has been started for you.
Glaciers as natural systems *(3.1.4.1)*				
4.1 Glaciated landscapes and the glacial system				*open system, inputs, outputs,*
The nature and distribution of cold environments *(3.1.4.2)*				
4.2 The nature and distribution of cold environments				
Systems and processes *(3.1.4.3)*				
4.3 Glacial systems and the glacial budget				
4.4 Historic patterns of ice advance and retreat				
4.5 Warm- and cold-based glaciers				
4.6 Geomorphological processes				
4.7 Periglacial features and processes				

Glaciated landscape development *(3.1.4.4)*				
4.8 Landscapes of glacial erosion				
4.9 Landscapes of glacial deposition				
4.10 Fluvioglacial processes and landforms				
4.11 Periglacial landforms				
Human impacts on cold environments *(3.1.4.5)*				
4.12 Fragility of cold environments				
4.13 Impacts of climate change on cold environments				
4.14 Management of cold environments				
Case studies *(3.1.4.7)*				
4.15 Challenges and opportunities for development – Svalbard, Norway				

Note: The 'missing' specification reference **3.1.4.6** refers to skills

You need to know:
- how a systems approach can be applied to a glacier
- that a glacial system can be linked to other natural systems.

Student Book
pages 158–9

What is a glaciated landscape?

Figure **1** is a distinctive glaciated landscape. The peaks surrounding the active glacier have been carved and shattered by ice processes. Rocks are dumped on the slopes by the melting ice or piled up beneath frost-shattered cliffs. Rock surfaces bear evidence of glacial erosion, with deep scratches or polished surfaces.

⊙ *Figure 1* *The Mer de Glace, France*

Systems terminology

Systems term	Examples/Processes
Inputs	Snow; avalanches; freeze-thaw debris
Outputs	Meltwater; icebergs (calving); evaporation; sublimation
Energy	Mass of glacier + gravity generates potential energy
Stores/ components	Snow; ice; debris
Flows/transfers	Meltwater flow; evaporation; sublimation; glacial movement
Feedback loops	Both positive and negative e.g. sediment on ice absorbs Sun's radiation and heats up, leading to melting and more sediment ... (+ve).
Dynamic equilibrium	Ablation (outputs) balanced by accumulation (inputs). The equilibrium line moves as the balance shifts.

⊙ *Figure 2* *Glacial systems terminology*

What is the glacial system?

A glacier can be viewed as an open system – with inputs from and outputs to external systems, such as atmospheric and fluvial systems.

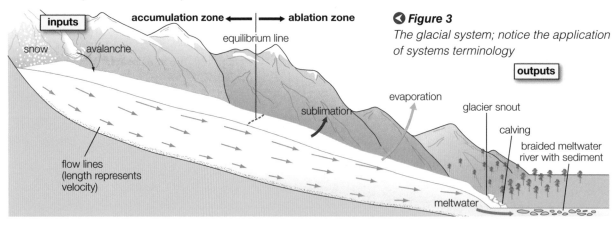

inputs

accumulation zone ◄——┃——► ablation zone

equilibrium line

snow   avalanche

⊙ *Figure 3*
The glacial system; notice the application of systems terminology

outputs

evaporation

sublimation

glacier snout

calving

braided meltwater river with sediment

flow lines (length represents velocity)

meltwater

Sixty second summary

- A glaciated landscape has distinctive characteristics including glaciers, steep mountainsides and deep U-shaped valleys (troughs).
- The glacial system is an open system, with inputs and outputs, processes and stores, positive and negative feedbacks and dynamic equilibrium.
- Glacial systems have clear links with other systems, such as atmospheric and fluvial systems.

Over to you

Make a list of the terms in the glacial system. Give a definition and an example for **each** term.

*Student Book
pages 160–3*

You need to know:

- the past and present-day distribution of cold environments
- the characteristics of cold environments
- about the nutrient cycle, demonstrating interactions between climate, soils and vegetation.

What was the past distribution of cold environments?

The geological period that lasted from about 1.8 million to 11 700 years ago is known as the Pleistocene. During this time there was a pattern of alternating cold periods (glacials) and warm periods (interglacials).

Figure **1** shows that much of Europe and North America was ice covered.

Figure **2** shows the extent of ice cover over the British Isles.

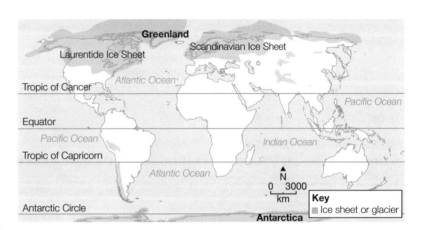

Figure 1 ◗
The extent of ice about 20 000 years ago, when ice covered large parts of Europe and North America

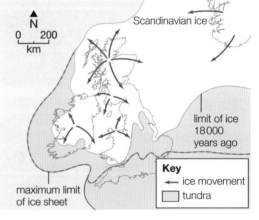

What is the present-day distribution of cold environments?

It is possible to identify four types of cold environment:

- *Polar* – areas of permanent ice; essentially Antarctica and Greenland.
- *Periglacial* (*tundra*) – permanently frozen ground (**permafrost**); large areas of northern Canada, Alaska, Scandinavia and Russia.
- *Alpine* – cold conditions due to altitude, particularly in winter; European Alps, Southern Alps in New Zealand.
- *Glacial* – edges of ice sheets and in mountainous regions; the Himalayas and Andes.

▲ **Figure 2** *The extent of ice cover about 18 000 years ago, when the British Isles was joined to Europe*

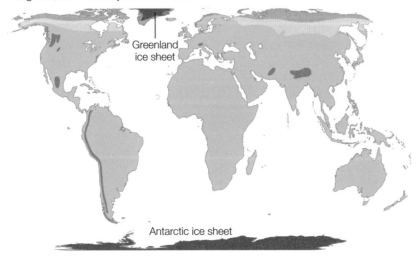

Key
- Polar
- Periglacial (continuous permafrost)
- Periglacial (discontinuous permafrost)
- Alpine

◗ **Figure 3**
This map shows the present-day global distribution of cold environments. Notice that almost all the cold environments are in the far north.

Continued over ▷▷▷

Physical characteristics of cold environments

Soil development

Soil formation is extremely slow, and any soils that do develop will be thin, acidic, sometimes waterlogged and mostly frozen (permafrost). In summer, the ground surface may thaw to form a boggy **active layer**.

Climate

- Significant periods when temperatures are close to, or significantly below, freezing. Liquid water is limited to certain times of the year or is totally absent.
- Varying amounts of snowfall – very little in polar environments to potentially huge amounts in coastal and mountain areas.
- Strong winds, which absorb precious moisture from plants.

Vegetation development

Vegetation is largely absent from polar and glacial environments, apart from at the very periphery where bare rock surfaces can be colonised by lichens.

Periglacial environments are characterised by tundra vegetation consisting of low-growing plants, including mosses, lichens, grasses, sedges and dwarf shrubs. Their small, waxy leaves are well adapted to retain warmth and reduce the water loss caused by exposure to strong winds. They maximise the short, warmer summers by flowering and setting seed in just a few weeks.

Climate, soils and vegetation interactions – nutrient cycling

Figure **4** shows the interactions in a cold environment between climate, soils and vegetation interactions. (This standard format of proportional circles and arrows is known as a Gersmehl diagram – see 6.4).

- Small nutrient stores indicating nutrient availability is very limited.
- Transfer arrows are generally thin indicating a very limited transfer of nutrients between components.
- The only sizeable transfer is the *fallout pathway* – organic matter such as dead leaves and animals, which contribute nutrients to the litter store.

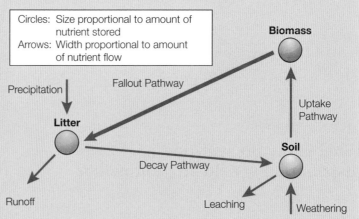

 Figure 4 *The nutrient cycle in a periglacial (tundra) environment. Notice how climate limits interaction between soils and vegetation*

- During past glacial periods, ice covered a large part of the Earth's surface particularly in Europe and North America.
- There are polar, periglacial, Alpine and glacial cold environments.
- Cold environments have distinctive climate, soils and vegetation.
- Lack of liquid water means slow soil formation, limited weathering and lack of vegetation and decomposers.
- Nutrient cycling demonstrates the links between climate, soils and vegetation – it is characterised by having small nutrient stores and slow/low transfers between them.

Use flashcards to summarise the past and present distribution of cold environments and their physical characteristics.

Student Book
pages 164–5

You need to know:

- explain what a glacial budget is (mass balance)
- understand its dynamic nature.

What is the glacial budget?

A financial budget has payments in (credits) and withdrawals (debits). In the **glacial budget** (mass balance), credits are inputs and withdrawals are outputs. Two zones can be identified (these are shown in Figure **3**, 4.1):

- The *accumulation zone*, where the inputs (e.g. snow, avalanches) exceed the outputs – a net gain of ice during the year.
- The *ablation zone*, where the losses (e.g. meltwater, calving) exceed the gains.

The boundary where gains and losses are balanced is the *equilibrium line*. Variations in the glacial budget may move the equilibrium line up or down the glacier. This can often be linked to the advance or retreat of the glacier snout.

In the summer, ablation will be at its highest due to rapid melting of the ice. During the winter, higher amounts of snowfall and limited melting will result in accumulation being greater than ablation.

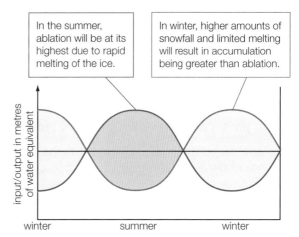

In the summer, ablation will be at its highest due to rapid melting of the ice.

In winter, higher amounts of snowfall and limited melting will result in accumulation being greater than ablation.

Figure 1 *This diagram shows the seasonal variations in the glacial budget; the red line is ablation, the blue line is accumulation*

Gulkana Glacier, Alaska, USA

Glaciers can be studied in order to understand glacier dynamics and hydrology, and assess the glaciers' response to climate change. One such 'benchmark' glacier is the Gulkana Glacier in Alaska. The climate is characterised by a large range in temperature, and low, irregular amounts of precipitation.

In studying fixed points on the glacier, together with meteorological and runoff data, United States Geological Survey (USGS) scientists have been able to both calculate and plot changing trends in the glacier's mass balance.

Data from the Gulkana case study can be examined on page 165 of the student book.

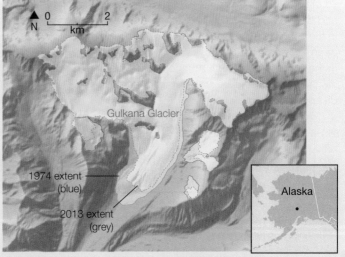

Figure 2 *Changes in extent of the Gulkana Glacier; 1974 extent in blue; 2013 extent in grey*

 Sixty second summary

- The glacial budget (mass balance) is the balance between the inputs (accumulation) and outputs (ablation).
- A glacier may be divided into the accumulation zone and ablation zone.
- The equilibrium line is where gains and losses are balanced.
- The Gulkana Glacier is used by the USGS to understand glacier dynamics, hydrology and assess glaciers' response to climate change.

 Over to you

Describe and explain how the glacial budget changes over the course of a year.

Student Book
pages 166–7

You need to know:

- the reasons why glaciers advance and retreat
- the historic patterns of glacier movement.

Why do glaciers advance and retreat?

Glaciers respond to long-term trends in the glacial budget, the balance between accumulation and ablation.

- If *accumulation* exceeds *ablation*, the glacier's mass increases and it will probably advance.
- If *ablation* exceeds *accumulation*, the glacier's mass decreases and it will probably retreat.

Historic patterns of ice advance and retreat

A recent cold period lasted from 1550 to 1850 – the 'Little Ice Age' – when global temperatures were cooler than the present day. Ice advanced, particularly in mountain areas. There were frost fairs on the frozen River Thames, winter food shortages and the formation of sea ice around the UK coastline.

Over the next hundred years the climate warmed and many of the world's glaciers retreated (Figure **1**). A slight global cooling between 1950 and 1980 halted this trend and some glaciers advanced. Since 1980 many of the world's glaciers have experienced considerable shrinkage, which is linked to climate change and global warming.

- In 2009, research by the University of Zurich found that 76 of 89 Swiss glaciers were retreating, 5 were stationary and 8 were advancing.
- In the Himalayas, an estimated 10% of the ice mass has been lost since the 1970s.
- In the Cascades in North America, all 47 glaciers being monitored by scientists are retreating; some have disappeared completely.

The Mer de Glace, France

During the Little Ice Age (1550 to 1850), the Mer de Glace extended to the floor of the Valle de Chamonix. Since then the glacier has retreated by 2300 m and thinned markedly. At the famous Montenvers Station, it has thinned by 150 m since 1820.

Retreat has not been continuous. During the 1970s and 1980s, global cooling led to the glacier advancing by 110 m.

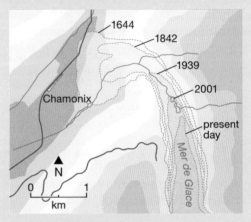

🔺 **Figure 2** *Past positions of the Mer de Glace, France*

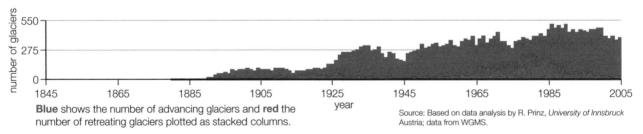

Blue shows the number of advancing glaciers and **red** the number of retreating glaciers plotted as stacked columns.

Source: Based on data analysis by R. Prinz, *University of Innsbruck Austria*; data from WGMS.

🔺 **Figure 1** *Worldwide changes in glacier length since the end of the 'Little Ice Age'. Notice the increase in number of advancing glaciers in the 1960s and 1970s.*

  Sixty second summary

- Glaciers advance and retreat in response to the balance between accumulation and ablation (the glacial budget).
- Historically, glaciers have advanced and retreated in response to natural cycles including the Little Ice Age.
- The recent retreat of many glaciers is the result of climate change associated with human impact (global warming).

Over to you

Use Figure **1** to identify and list key trends (time periods of glacial advance and retreat).

Student Book
pages 168–9

You need to know:
- definitions and characteristics of warm- and cold-based glaciers
- the concept of pressure melting point and its impact on movement and processes.

Warm-based glaciers

The melting point of ice (the *pressure melting point* (pmp)) decreases with increased pressure. If the temperature of the ice is the same as the pmp, melting will take place.

Figure **1** shows the temperature profile of a **warm-based glacier**. They are associated with a temperate environment, e.g. the European Alps.

- When temperatures exceed 0°C at the surface, melting will occur.
- At the glacier's base, the temperature of the ice reaches pmp producing meltwater allowing **basal sliding** (see 4.6).
- This results in significant erosion and deposition of large volumes of sediment.

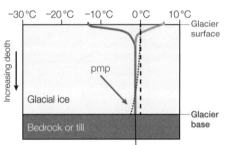

Temperature at base is **higher** than pmp so melting will occur and this allows basal sliding.

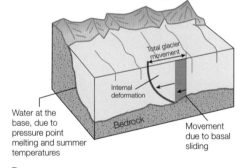

Water at the base, due to pressure point melting and summer temperatures

Movement due to basal sliding

⬆ **Figure 1** *Temperature profile for a warm-based glacier*

Cold-based glaciers

Figure **2** shows the temperature profile of a **cold-based glacier**. They are associated with polar environments, e.g. Antarctica or central Greenland.

- Temperatures are below 0°C all year round.
- The ice never reaches the pmp and remains frozen to the bedrock – basal sliding will not occur.
- Any movement will be solely by **internal deformation** (see 4.6).
- As a result, there will be little erosion or deposition.

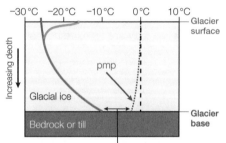

Temperature at the base is **well below** pmp so no melting at base and no basal sliding.

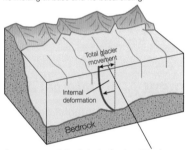

Far less movement due to lack of water at the base, so more friction. Ice moves by internal movement only.

⬆ **Figure 2** *Temperature profile for a cold-based glacier*

Key
— Average summer surface temperatures
— Average winter surface temperatures
— Average annual temperature
········· Pressure melting point (pmp) (the melting point of ice when under pressure)

Sixty second summary

- Warm-based glaciers are found in temperate regions; cold-based glaciers in polar regions.
- Pressure lowers the melting point of ice (pressure melting point), so ice melts beneath warm-based glaciers, assisting movement (basal slip).
- Cold-based glaciers have little, if any, meltwater so remain frozen to the bedrock. Limited movement (by internal deformation) results.
- Meltwater is important in creating depositional landforms.

Over to you

Write clear definitions of pressure melting point, warm-based glaciers and cold-based glaciers.

Student Book
See pages 170–5

You need to know about:

- the processes of weathering and nivation in cold environments
- ice movement mechanisms and factors affecting rate of ice movement
- glacial processes of erosion, transportation and deposition.

Weathering in a cold environment

Frost shattering

Frost shattering (freeze-thaw) commonly affects bare rocky outcrops high up on a mountainside.

- Water seeps into cracks and holes (pores) within a rock.
- The water freezes and expands.
- With repeated freezing and thawing, the cracks are enlarged until rock breaks away and piles up as scree at the foot of the slope.
- Frost-shattered rocks are sharp and angular. When trapped under the ice they are an extremely effective abrasive tool.

Carbonation

Carbon dioxide dissolved in water forms a weak carbonic acid. This reacts with calcium carbonate in some rocks, particularly limestone, to form calcium bicarbonate. This is *carbonation* – a process of chemical weathering. Carbon dioxide is more soluble at low temperatures which is why it is an important process in cold environments.

Nivation

Nivation is an umbrella term used to cover a range of processes most active around the edges of snow patches (Figure **1**).

- Fluctuating temperatures and meltwater promote frost shattering.
- Summer meltwater carries away weathered material enlarging the *nivation hollow*.
- Slumping may also take place during the summer, as saturated debris collapses due to the force of gravity.

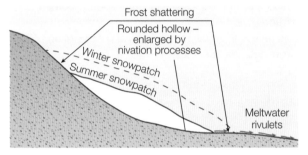

🔺 **Figure 1** *Nivation processes result in the formation of a nivation hollow*

Ice movement

Basal sliding

This involves movement (2–3 m a day) where meltwater lubricates the base of the ice, transporting debris which erodes underlying bedrock. When a glacier encounters an outcrop of tough rock *regelation* takes place:

- Increased resistance (and therefore pressure) on the upslope side, causing melting of the base – *pressure melting*.
- This eases movement over the obstacle.
- Pressure reduced downslope, so meltwater refreezes.

Internal deformation

Internal deformation takes place in both cold- and warm-based glaciers (1–2 cm a day). There are two mechanisms at work:

- *Intergranular movement*: individual ice crystals slip and slide over each other.
- *Intragranular movement*: individual ice crystals become deformed or fractured due to stresses within the ice.

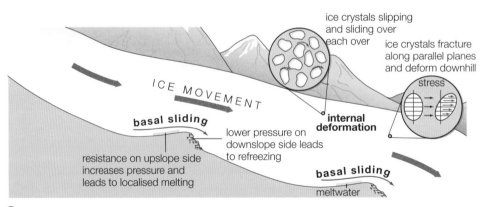

🔺 **Figure 2** *This diagram shows basal sliding and internal deformation*

Variations in the rate of ice movement

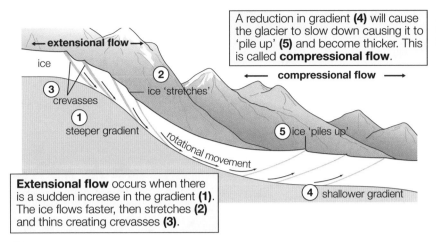

A reduction in gradient **(4)** will cause the glacier to slow down causing it to 'pile up' **(5)** and become thicker. This is called **compressional flow**.

Extensional flow occurs when there is a sudden increase in the gradient **(1)**. The ice flows faster, then stretches **(2)** and thins creating crevasses **(3)**.

⬥ **Figure 3** *Illustration of extensional, compressional and rotational flow*

A negative feedback loop could take place.

Thickening of ice (**compressional flow**) increases mass and potential erosion. This could lead to a steeper gradient and thus encourage faster **extensional flow**, a thinning of the ice and a reduction in potential erosion.

Controlling factors in ice movement

- *Gravity* – the steeper the gradient, the greater the pull of gravity.
- *Friction* – friction exerted on the ice has to be overcome.
- *Mass of the ice* – the heavier the ice, the more potential energy it has to move.
- *Meltwater* – meltwater lubricates the base of the ice.
- *Temperature of the ice* – in some environments the ice is so cold that it is frozen to the bedrock.

Glacial processes

Glacial erosion

There are two main forms – abrasion and plucking.

- *Abrasion* is the sandpapering effect of ice as it grinds over and scours a landscape (Figure **4**). Rocks carried beneath the ice will often scratch the bedrock (*striations*).
- *Plucking* or *quarrying* occurs when meltwater freezes part of the underlying bedrock to the base of a glacier. Any loosened rock fragments will be 'plucked' away as the glacier subsequently slips forward.

Glacial ice, up to 2 km thick

Loose rocks are plucked from the glacier as it moves forward

Direction of ice flow

Sandpapering effect of abrasion

Bedrock

 Figure 4 *The formation of a roche moutonnée by abrasion and plucking processes*

Glacial transportation

Glaciers transport material in three ways:

- *Supraglacial* – predominantly weathered material carried on top of the ice.
- *Englacial* – supraglacial material, buried by snowfall, and carried within the ice.
- *Subglacial* – material dragged beneath the ice by the overlying glacier.

Glacial deposition

Deposition of sediment transported by the ice takes place when the ice melts, which is primarily in the ablation zone close to the glacier's snout. Sediment on or in the ice melts out in the same way the features of a snowman are left on the ground when it melts. Water can carry sediment many kilometres from the ice.

Sixty second summary

- Major active weathering processes are associated with ice formation such as frost shattering.
- Nivation is a group of processes associated with snow (e.g. weathering and mass movement) which combine to form nivation hollows.
- Ice moves by basal slip or internal deformation, responding to factors such as temperature of the ice, gravity and meltwater.
- Glacial processes include erosion (through abrasion and plucking), transportation and deposition.

Over to you

Summarise key points about geomorphological processes operating in cold environments (weathering, ice movement and glacial processes).

You need to know:
- the characteristic features of periglacial environments
- processes operating in periglacial environments, including frost action and mass movement
- why permafrost varies in depth and extent.

Features of a periglacial environment

Permafrost is where a layer of soil, sediment or rock below the ground surface remains frozen for more than a year. During the warm summers, the top part melts, forming a potentially mobile **active layer** (Figure **1**).

Thaw lakes are common in poorly drained periglacial areas. Water absorbs radiation from the Sun and retains it. This increases the depth at which the underlying permafrost melts, forming unfrozen zones called *taliks* (Figure **1**).

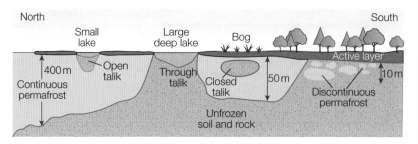

▲ **Figure 1** *Cross-section across permafrost in the Arctic. Notice that the active layer thickens and the permafrost becomes less continuous further to the south, away from the Arctic.*

Processes in a periglacial environment

Frost action

Frost action is most commonly associated with the weathering process of *frost shattering* (*freeze–thaw*). This results in scree at the foot of a slope or a boulder-strewn landscape called **blockfield** or *felsenmeer*.

Frost action can also be effective through *ground ice*. Ice can form within pores (*pore ice*) or as *ice needles*, which can force individual soil particles or small stones to the surface (**frost heave**).

Cracks can fill with water in the summer, which freezes and enlarges the crack. Over many years an **ice wedge** is formed, which can be 3 m wide at the surface and 10 m deep (Figure **2**).

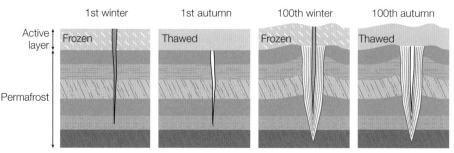

▲ **Figure 2** *The formation of ice wedges*

Mass movement

Solifluction is when saturated soil (the active layer) slumps downhill during the summer to form solifluction lobes.

Frost creep is particles being raised perpendicular to the ground surface by frost heave. They are then dropped down vertically (due to gravity) on thawing. Repeated cycles cause soil particles to gradually 'creep' downslope.

Sixty second summary

- Periglacial environments exhibit distinctive features, such as permafrost and thaw lakes.
- Many processes in perigiacial environments are associated with ice action, such as frost shattering and the formation of ice wedges.
- Mass movement is active particularly during the summer thaw (e.g. solifluction, frost creep).

Over to you

Create simple spider diagrams to outline the features of a periglacial environment and the processes operating in a periglacial environment.

Student Book
pages 178–81

You need to know:
- the characteristics and formation of erosional landforms within the glaciated landscape
- the processes involved in their development.

What is a glacial erosion landscape?

Figure **1** shows the iconic landscape of the Lake District –a typical glaciated landscape.

Red Tarn – a corrie lake

Corrie – a scooped out hollow in the landscape

Striding Edge – a classic example of an arête

Glacial troughs or U-shaped valleys.

Figure 1 *Photo from Helvellyn, showing landforms of glacial erosion*

The origin and development of glacial erosional landforms

Corries

A *corrie* is an enlarged hollow on a mountainside. Figure **2** shows how a corrie is formed over time.

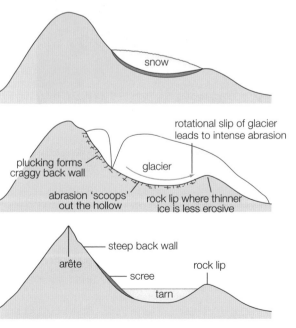

snow

rotational slip of glacier leads to intense abrasion

plucking forms craggy back wall

glacier

abrasion 'scoops' out the hollow

rock lip where thinner ice is less erosive

arête

steep back wall

scree

rock lip

tarn

Figure 2 *These diagrams show how a corrie is formed*

a) Periglacial conditions: periglacial processes (nivation, frost action) increase the size of a hollow or depression on the mountainside.

b) Glacial conditions: as the climate cools, snow turns to ice and a glacier develops. Accumulation increases its mass and rotational sliding 'scoops out' of the hollow by abrasion.

c) Post-glacial conditions: as the climate warms, periglacial and then temperate processes (including frost and water action) modify the corrie to create what we see today.

Arêtes and pyramidal peaks

An *arête* is a knife-edge ridge formed when two neighbouring corries cut back into a mountainside. Where three or more corries erode back-to-back, the ridge becomes *a pyramidal peak* (Figure **3**).

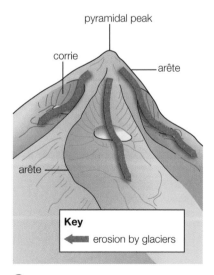

pyramidal peak

corrie

arête

arête

Key
← erosion by glaciers

Figure 3 *How arêtes and pyramidal peaks are formed*

Continued over ⟩⟩⟩

Glacial troughs

Glacial troughs are steep-sided, broadly flat-bottomed and often several hundred metres in depth. They tend to be straight because of the inflexibility and immense power of the glaciers that gouge them out.

Glacial troughs may contain deep, narrow lakes called *ribbon lakes*. They result from localised overdeepening due to:

- weaker bedrock allowing increased vertical erosion
- a merging tributary glacier adding to the mass of ice and increasing erosion of the valley
- narrowing of the valley and subsequent thicker ice, leading to increased vertical erosion.

Hanging valleys

A glacier occupying a tributary valley does not have the same mass as a larger glacier in a trunk valley; it cannot erode down as far as the larger glacier. When the ice melts, the smaller valley is left 'hanging' high above the main valley – a *hanging valley* (Figure **4**).

Truncated spurs

When a valley is occupied by ice, the rigid glacier cuts off the tips of the interlocking spurs, by the processes of abrasion and/or plucking. This leaves behind steep cliffs – *truncated spurs*.

Roche moutonnées

A *roche moutonnée* is a bare outcrop of rock on the valley floor that has been sculpted by the moving ice (Figure **4**, 4.6).

- On the upstream side, an increase in pressure caused by the resistance of the rock outcrop to the moving ice, leads to localised pressure melting. This facilitates basal sliding and the process of abrasion (causing striations) as the glacier slides over the rock outcrop.
- On the downstream side, the reduction in pressure causes the meltwater to freeze forming a bond between the rocky outcrop and the overlying glacier. As the glacier continues to move forward it plucks away loose rocks leaving behind a jagged surface.

See page 181 of the student book for an Ordnance Survey map of a classic landscape of glacial erosion.

Controlling factors in the extent of erosion

- *Mass of the ice* – the thicker the ice, the greater potential energy for erosion.
- *Gradient* – this can affect the rate of flow of the ice and also its thickness (*extensional* and *compressional flow*, see 4.6).
- *Meltwater* – enables basal sliding to occur, which is likely to be more erosive than internal deformation.
- *Rock debris* – rocks trapped beneath the ice that scrape and gouge the underlying bedrock.
- *Underlying geology* – whether the rocks are strong or weak, massive or thinly layered, jointed or unjointed.

Hanging valley of former tributary glacier

Truncated spur

Valley of former main glacier

 Figure 4 *Stirling Falls in New Zealand – a hanging valley*

Sixty second summary

- A corrie is a deep mountain hollow formed by a combination of erosional processes together with weathering and mass movement.
- Arêtes and pyramidal peaks are created when multiple corries meet.
- Glacial troughs illustrate the enormous erosive power of glaciers, particularly abrasion through basal sliding.
- Hanging valleys, truncated spurs and roche moutonnées are erosional landforms associated with glacial troughs.

Over to you

Use a series of flashcards to outline the characteristics and formation of glacial erosional landforms.

Student Book
pages 182–5

You need to know:

- the characteristics of sediment deposited by glaciers
- the origin and development of landforms associated with glacial deposition.

What is a landscape of glacial deposition?

Figure **1** shows a landscape with features that have been formed by glacial deposition. These landscapes tend to have the following characteristics:

- Glaciers transport rock debris from eroded mountains and deposit it on valley floors or lowland plains.
- Rock debris dumped in situ when the ice melts is angular and poorly sorted (*till*).
- Meltwater streams can carry sediment many kilometres, depositing it as well-sorted **outwash plains**.

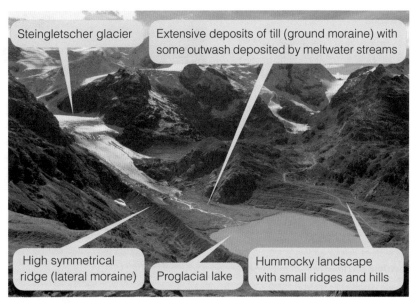

Steingletscher glacier

Extensive deposits of till (ground moraine) with some outwash deposited by meltwater streams

High symmetrical ridge (lateral moraine)

Proglacial lake

Hummocky landscape with small ridges and hills

▲ **Figure 1** *Photo of the Steingletscher glacier in the Swiss Alps, a landscape of glacial deposition*

Landforms of glacial deposition

Moraines

Moraine is a generic term for landforms associated with the deposition of till. It is therefore poorly sorted with predominantly angular sediments. There are several types:

- *Ground moraine* – sediment transported beneath a glacier that is smeared over the underlying bedrock.
- *Terminal moraine* – a ridge of sediment piled up in front of the glacier. It marks the furthest extent of an advancing glacier.
- *Recessional moraine* – a secondary ridge formed at the snout of a retreating glacier during periods of stability.
- *Lateral moraine* – a ridge formed alongside a glacier primarily from the build-up of scree slopes.
- *Medial moraine* – when two glaciers merge, lateral moraines at the edges of the two glaciers join to form a medial moraine.

a) Lowland landscape during glaciation

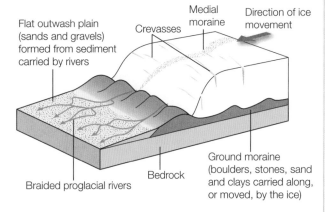

Flat outwash plain (sands and gravels) formed from sediment carried by rivers

Crevasses

Medial moraine

Direction of ice movement

Braided proglacial rivers

Bedrock

Ground moraine (boulders, stones, sand and clays carried along, or moved, by the ice)

b) The same lowland landscape after glaciation

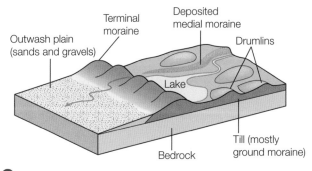

Outwash plain (sands and gravels)

Terminal moraine

Deposited medial moraine

Drumlins

Lake

Bedrock

Till (mostly ground moraine)

▲ **Figure 2** *Landforms of glacial deposition*

Continued over ▶▶▶

Till plain

A *till plain* is an extensive plain resulting from the melting of a large sheet of ice that became detached from a glacier. This has the effect of levelling out the existing landscape.

Erratics

An *erratic* is most commonly a boulder or a smaller rock fragment that has been deposited in a location that is foreign to its origin (Figure **3**). They give an indication of the direction of ice movement – find where the erratic came from and you have the direction of ice flow.

Drumlins

A typical *drumlin* is an oval- or egg-shaped hill composed of glacial till and aligned in the direction of ice flow (Figure **4**). They tend to be 30–50 m in height and up to 1 km in length, and usually occur in clusters or 'swarms' on flat valley floors or lowland plains. Drumlins are common in parts of northern England, Scotland and Canada.

Some drumlins have a rocky core, almost most do not. Some consist of fluvial sediment as well as glacial till, which suggests that meltwater plays an important role in their formation, and may be a combination of several processes.

The difficulty in determining exactly how they are formed comes from their formation being beneath the ice, so observation is not possible.

 The formation of drumlins are not fully understood. Page 184 of the student book looks at two theories.

Figure 3 *An erratic boulder in the Yorkshire Dales. You can see that the erratic is clearly a different rock type than the limestone it is sitting on.*

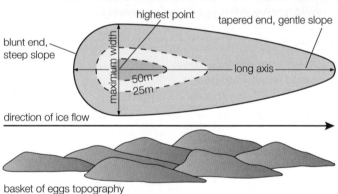

basket of eggs topography

Figure 4 *Plan view of a drumlin and the 'basket of eggs' topography of a drumlin field. The cross-section is aligned with the photo to show the steep and tapered slopes*

Sixty second summary

- Glaciers transport huge amounts of sediment, called till.
- Till is either unsorted when deposited by the glacier or well sorted when carried away from the glacier by meltwater forming outwash plains.
- Landforms of deposition include moraines, till plains, erratics and drumlins.
- Drumlins and erratics give clues to the direction of flow of the ice.

Over to you

You need to be very confident of the terminology used throughout this topic. Summarise the depositional features and formations outlined using a poster, flashcards, spider diagram or whatever works best for you. Take care over clarifying the different types of moraine.

Student Book
pages 186–9

You need to know:

- the characteristics of fluvioglacial landscapes
- processes operating in fluvioglacial landscapes
- which landforms are associated with fluvioglacial processes and their formation.

What are fluvioglacial landscapes?

Fluvioglacial landscapes are associated with flowing water – essentially meltwater – in glacial or periglacial environments.

- Meltwater is seasonally abundant in temperate glacial and periglacial environments.
- It is often seen at a glacial snout flowing out from under the ice (Figure **1**).
- It is much less common in very cold environments, characterised by *cold-based glaciers*.
- A fluvioglacial landscape is very dynamic, with river channels constantly changing course.

Figure 1 *Meltwater flowing from beneath a glacier in Prince William Sound, Alaska*

What are fluvioglacial processes?

Meltwater plays a crucial role in several glacial and periglacial processes.

- *Nivation* – meltwater removes broken rock during the summer when the outer edges of the snow patch melts. It is also vital for the process of freeze–thaw.
- *Basal sliding* – meltwater lubricates the base of *warm-based glaciers*.
- *Abrasion* – meltwater beneath a glacier provides rocks that are used as 'tools' for erosion by abrasion.
- *Plucking* – meltwater refreezes to rock fragments, 'plucking' rocks.
- *Depositional features* – meltwater erodes channels both beneath the ice and in front of it, forming distinctive features characterised by well-sorted, rounded, smoothed deposits.

What are the distinctive fluvioglacial landforms?

Meltwater channel

Overspill from a lake alongside or in front of a glacier carves a steep-sided **meltwater channel**, which are now often 'dry valleys', e.g. Newtondale in North Yorkshire (Figure **2**). Today, this is a narrow, wooded gorge some 80 m deep and 5 km in length.

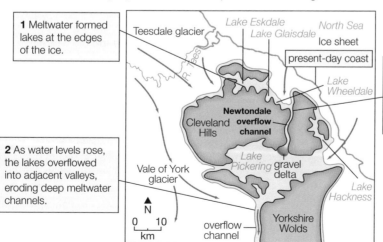

1 Meltwater formed lakes at the edges of the ice.

2 As water levels rose, the lakes overflowed into adjacent valleys, eroding deep meltwater channels.

3 Newtondale was formed when water overflowed from Lake Wheeldale southwards towards Lake Pickering.

Figure 2 *The formation of Newtondale meltwater channel*

Continued over ▶▶▶

Outwash plain

An **outwash plain** is an extensive, gently sloping area in front of a glacier resulting from the 'outwash' of sand and gravels carried by meltwater. Some of the most extensive outwash plains are in Iceland and Alaska.

Eskers

Eskers are long, sinuous (winding) ridges made of sand and gravel that can be up to 30m high and stretch for several kilometres (Figures **3**). They usually take the form of meandering hills running roughly parallel to the valley sides. They are formed by *subglacial* river deposition during the final stages of a glacial period, when the ice is in recession (Figure **4**).

Kames

As with eskers, **kames** are also largely made of sand and gravel, but they are deposited *on the surface* of the ice by streams in the final stages of a glacial period. There are different types of kame:

* *Kame terrace* – results from the infilling of a marginal glacial lake. When the ice melts, the kame terrace is abandoned as a ridge on the valley side.
* *Kame delta* – forms when a stream deposits material on entering a marginal lake. They form small, mound-like hills on the valley floor.
* *Crevasse kame* – a result of the fluvial deposition of sediments in surface crevasses. When the ice melts they are deposited on the valley floor to form small hummocks.

 Figure 3 *The Dahlen esker, North Dakota; one of the best examples of an esker in the USA*

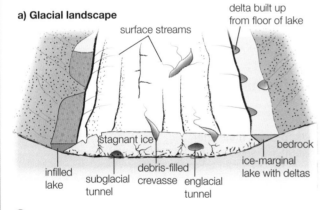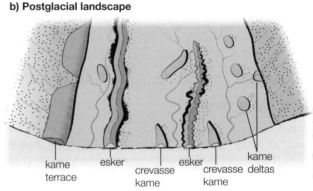

a) **Glacial landscape**

surface streams
delta built up from floor of lake
stagnant ice
bedrock
infilled lake
subglacial tunnel
debris-filled crevasse
englacial tunnel
ice-marginal lake with deltas

b) **Postglacial landscape**

kame terrace
esker
crevasse kame
esker
crevasse kame
kame deltas

Figure 4 *Eskers are formed from deposits **beneath** the ice; kames from deposits **on top** of the ice*

Sixty second summary

* Fluvioglacial landscapes are associated with meltwater in glacial and periglacial environments.
* Meltwater is seasonally abundant in temperate glacial and periglacial environments.
* Meltwater plays an important role in glacial movement and in glacial processes.
* Meltwater is responsible for the formation of distinctive fluvioglacial landforms (e.g. meltwater channels, outwash plains, eskers, kames).

Over to you

Review both 4.9 and this spread again. Create a simple summary table to contrast the key distinguishing characteristics of till and fluvioglacial deposits.

Student Book
pages 190–3

You need to know:

- the characteristics of periglacial landscapes
- the characteristics and formation of distinctive periglacial landforms.

What is a periglacial landscape?

Periglacial landscapes, such as in Alaska and parts of Northern Europe, Russia and Canada, are wide expanses of largely featureless plains, either strewn with rocks (*blockfields/felsenmeer*) or clothed with low-growing, marshy vegetation. Lakes or streams are common in the summer when the snow has melted and some of the permafrost has thawed.

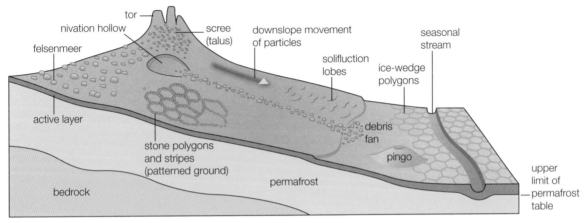

▲ **Figure 1** *Typical landforms associated with a periglacial environment*

What are the distinctive periglacial landforms?

Permafrost

Permafrost is soil, rock or sediment that has been frozen for at least two consecutive years – a characteristic feature of a periglacial environment. Many landforms owe their formation, at least in part, to the presence of permafrost and the action of ice (Figure **1**).

Ice wedges

In extremely low temperatures, the ground contracts and cracks develop. During the summer, meltwater fills these cracks and then freezes in the winter to form *ice wedges*, which increase in size through repeated cycles of freezing and thawing (Figure **2**).

Patterned ground

As ice wedges become more extensive, a polygonal pattern may be formed on the ground, with the ice wedges marking the sides of the polygons. This is **patterned ground**.

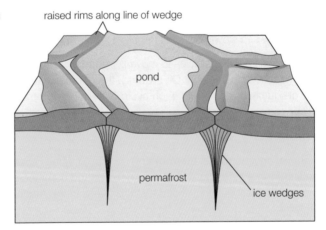

▲ **Figure 2** *Surface features associated with ice wedges; notice the raised edges that trap meltwater*

Stone polygons form on flat ground and are directly associated with ice wedges. *Frost heave* causes expansion of the ground and lifts soil particles upwards. Smaller particles are removed by wind or meltwater, leaving larger stones lying on top of the ice wedges marking out the polygonal pattern.

Sloping ground causes stones to move downhill under gravity, leading to the formation of elongated *stone stripes* rather than polygons (Figure **1**).

Continued over ❯❯❯

4.11 **Periglacial landforms**

Blockfields

Periglacial landscapes in mountainous regions, can be characterised by extensive frost-shattered bedrock consisting of broken up angular fragments of rock – *blockfields* or *felsenmeer* (Figure **1**). These areas are subject to repeated cycles of freezing and thawing.

Pingos

Pingos are one of the most spectacular periglacial landforms found in the Arctic and sub-Arctic (Figures **1** and **3**). They can reach heights of up to 90m. They are often green and vegetated on the outside, but their core is solid ice.

Formation of a 'closed' pingo

Stage 1

- Lake infills with sediment, insulating the ground beneath.
- Liquid water is trapped in the unfrozen ground (talik) between the lake sediment and permafrost.

Stage 2

- Water freezes as climate cools to form ice core.
- Ice expands due to increased hydrostatic pressure and the talik is squeezed.
- Lake sediment pushed up to form a pingo. In summer, part of the ice core may collapse and fill with water.

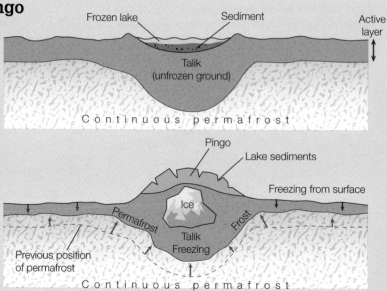

▲ **Figure 3** *This diagram shows how a 'closed' pingo is formed on the site of a lake. 'Closed' pingos are typical of Northern Canada.*

Solifluction lobes and terracettes

Solifluction involves the downslope movement of rock and soil material due to gravity. This results in tongue-like landforms called *solifluction lobes*. In the summer, the saturated *active layer* will simply slip slowly downhill (Figure **1**).

Frost heave causes soil particles to move perpendicular towards the surface due to expansion caused by freezing. On thawing, the particles fall back vertically, causing particles to move gradually downhill with each cycle of freezing and thawing. On some slopes, distinct 'steps' can form – **terracettes**.

Thermokarst

In periglacial environments, **thermokarst** can result when ground ice melts and settles unevenly to form a landscape of marshy hollows and hummocks. This type of landscape is most commonly associated with flat lowland plains in the Arctic.

Sixty second summary

- Periglacial landscapes are characterised by large expanses of open land with rock exposures and low-growing, marshy vegetation.
- Periglacial landforms owe their formation, at least in part, to the presence of permafrost and the action of ice.
- Ice wedges can create low ridges enclosing thaw ponds and patterned ground (polygons and stripes).
- Blockfields are the result of intensive frost shattering.
- Pingos, solifluction lobes/terracettes and thermokarst are common landforms in periglacial regions associated with seasonal climatic variability.

Over to you

Practise drawing simple, annotated sketches of periglacial landforms in order to summarise their characteristics and formation.

Student Book
pages 194–5

You need to know:

- the reasons for environmental fragility in cold environments
- examples of fragility – oil spills in Siberia and tourism in the European Alps.

Why are cold environments fragile?

There are several reasons why cold environments are fragile:

- Slow ecosystem development and highly specialised habitats – plants and animals have adapted to the lack of daylight, harsh climatic conditions and the short growing seasons.
- Plants and animals that have adapted to very particular environmental conditions are very sensitive to change.
- Once damaged, an ecosystem can take a very long time to recover or it might never recover. Oil spills are not uncommon, especially in Russia (Figure **1**).

The cause of the spill was a decommissioned oil well

Thick, toxic, inflammable crude oil formed lakes on the tundra and polluted rivers

⚠ *Figure 1 Ecological catastrophe at Usinsk, near the Arctic Circle*

Human impacts on fragile cold environments?

Oil spills

The Arctic holds extensive and highly valuable reserves of oil and natural gas (Figure **2**) and there is huge potential for exploitation, but also for environmental damage. Oil spills contaminates the soil, kills all plants that grow in it and destroys habitats for mammals and birds (Figure **1**).

Tourism in alpine environments

Tourism in the European Alps has increased significantly in recent years due to higher levels of wealth and increased mobility and leisure time.

The most destructive human impacts are associated with the lucrative ski industry, which has had scant regard for environmental conservation. Forests have been stripped to make way for ski developments and infrastructure, resulting in habitat loss or fragmentation.

Modern adventure sports and some motor-based leisure activities are causing a major disturbance to wildlife in the Alps.

⚠ *Figure 2 Over 87% of the Arctic's oil and natural gas resource is located in these seven huge basins*

Sixty second summary

- Cold environments are fragile because of slow ecosystem development, highly specialised adaptations and high sensitivity to change.
- Oil spills damage terrestrial and aquatic ecosystems and can take a long time to recover, if at all.
- Tourism developments, e.g. ski lifts, transport infrastructure, have damaged Alpine ecosystems.

Over to you

Consider additional annotations to Figure **1** outlining the specific risks of economic development in cold environments highlighted by this oil spill.

You need to know:
- the impacts of climate change on natural systems in cold environments
- the impacts of climate change on human activity.

Student Book
pages 196–9

How is climate change affecting cold environments?

Climate change in cold environments – particularly in the Arctic – threatens to have a very significant impact on natural ecosystems and habitats. Figure **1** shows that the average temperature in the Arctic has increased by more than 2 °C since 1960.

Read about positive feedback loops (Arctic amplification) on page 197 of the student book.

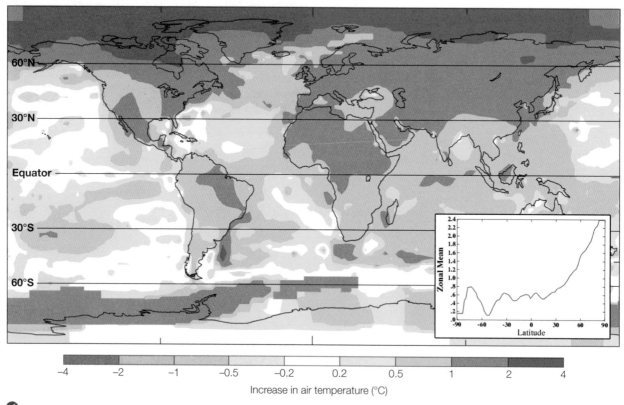

Increase in air temperature (°C)

◀ **Figure 1** *Map showing trends in mean surface air temperature (1960–2011)*

Impact of climate change in the Swiss Alps

- *Flooding and landslides*: in Switzerland almost all glaciers are retreating. In the summer, significant melting occurs and water can be dammed behind debris and released suddenly as powerful torrents of water.

- *Melting permafrost*: permafrost covers about 5% of Switzerland. Predicted thawing has increased the threat of rock avalanches and mudslides.

- *Ecosystems*: a rise in temperature of 1 °C pushes the treeline up by 100 m. If this happens, some of the specialist high-mountain flora and fauna will become extinct; habitats will change, and some species such as marmot will be threatened.

- *Economic costs*: a study by the Swiss National Research Programme has estimated that a rise of 2 °C will cost Switzerland about CHF 3 billion (£2 billion) a year, mostly affecting winter tourism. The study also said that expensive measures would have to be implemented to reduce the potential impact of flooding and other natural disasters.

- *Ski industry*: research by the Swiss National Science Foundation suggests that by 2050 only those resorts at 1500 m and above (62% of all resorts) will be able to offer skiing for at least a hundred days a year. Several low-altitude resorts already suffer from declining tourist numbers, with local businesses being forced to close.

Impact of climate change in the Arctic

Sea ice thickness and extent

The warming of the Arctic has resulted in a reduction in the thickness and extent of the Arctic sea ice.

- Satellite data show that over the past 30 years, Arctic sea ice cover has declined by 30% (Figure **2**).
- Satellite data also show that snow cover over land in the Arctic has decreased, and glaciers in Greenland and northern Canada are retreating.
- Permafrost in parts of the Arctic has started to thaw.

Indigenous population

For over a thousand years in November, the nomadic Nenet tribes have migrated across the frozen Ob River on the Yamal peninsula in north-west Siberia. Increased temperatures are delaying the freezing of the river, preventing tribes from crossing until December. This delay means the reindeer go hungry due to the lack of pasture.

In the high Arctic the reduction in sea ice, together with rising sea levels, is making coastal communities more vulnerable to flooding from high waves and storm surges. However, one positive impact is that harbours are ice-free for longer periods, giving more opportunities for fishing.

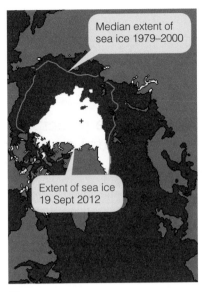

Median extent of sea ice 1979–2000

Extent of sea ice 19 Sept 2012

▲ **Figure 2** *The white area is the sea ice extent on 19/9/12 – the smallest extent since records began*

Impact of climate change on ecosystems

Climate change – in particular a warming trend – could have significant impacts on ecosystems.

Sea ice ecosystems

- Polar bears are dependent on sea ice for their entire lifecycle – from hunting seals, to raising their cubs.
- An increasing number of polar bears are drowning because they have to swim longer distances between ice floes.
- The lack of available food has resulted in polar bears fasting on land or even cannibalising each other.
- Similarly, ice-dependent seals give birth and nurse their pups on the ice and use it as a resting platform. It is very unlikely that these species could adapt to life on land in the absence of summer sea ice.

Land ecosystems

- Rapid environmental change associated with global warming, such as earlier springs, extended growing seasons and changes to nutrient availability, may occur quicker than the ability of Arctic plants to adapt.
- Some animals, such as musk oxen and reindeer, depend on snow cover to insulate plants. With reduced snow, plants may be less able to survive the winter, which impacts on the animals' food supply.
- Insect pollination in the Arctic has evolved over a long period of time. If plants start to flower earlier but changes to insect life cycles do not happen at the same rate, pollination will be less successful.

Sixty second summary

- The average temperature in the Arctic has increased by more than 2 °C since 1960.
- Climate change is affecting natural ecosystems and is also having economic and social impacts on people.
- Climate change is increasing the risk of avalanches and landslides in the Swiss Alps, reducing snow reliability and so, affecting tourism and economic activities.
- Climate change has reduced the thickness and extent of Arctic sea ice, affecting the way of life of indigenous people and impacting fragile ecosystems.

Over to you

For **each** of the Swiss Alps and the Arctic, summarise the key impacts of climate change.

You need to know:

- about management issues involving oil exploitation in Alaska
- about management issues associated with developments in the Alps.

Student Book
pages 200–3

What are the principles of environmental management?

It is possible to identify three management approaches:

1 *Prevention* – attempting to prevent a harmful event occurring, e.g. deforestation.
2 *Reaction* – responding to an event once it has occurred, e.g. clearing an oil spill.
3 *Adaptation* – learning to live with change, e.g. climate change.

Adapting to climate change in the European Alps

A report funded by the European Union, *Climate Change and its Impacts on Tourism in the Alps* (2011), has identified opportunities and threats associated with climate change.

- Summer tourism could benefit from climate change. Hotter summers (as in 2003) would bring more people to the mountains and the tourism season could be extended.
- Winter tourism faces serious challenges due to the expected decrease in snow and ice cover. Already, 57 of the main ski resorts in the European Alps are considered not to be snow-reliable.
- Many resorts use artificial snow, which not only use huge amounts of water and energy, but also may have detrimental impacts on ecosystems.
- Droughts may become more frequent in the summer.

Pages 202 and 203 of the student book have examples of how places in the Alps are adapting to climate change.

Sixty second summary

- Oil exploration in Alaska faces many challenges – the Trans-Alaska pipeline is considered to be a successful management response.
- In the Alps, hotter summers could extend the tourist season, but many winter ski resorts are already no longer snow-reliable.
- Responses include higher altitude skiing facilities, use of water and energy-demanding snow cannons, and protective measures to reduce the risk of avalanches and landslides.

Big idea

Management of cold environments involves balancing economic demands with environmental considerations.

Oil exploitation in Alaska, USA

The Prudhoe Bay oilfield is the largest oilfield in North America (Figure **1**). It has produced a significant proportion of the USA's energy needs for four decades. In a world of increasing energy insecurity, the Alaskan reserves are extremely important to the US both politically and economically.

The Trans-Alaska pipeline – successful management?

The most ambitious project in the region has been the 1287 km long Trans-Alaska oil pipeline. The insulated pipeline runs from Prudhoe Bay in the north to the ice-free port of Valdez in the south. The pipeline is raised on stilts, to prevent the warm oil melting the underlying permafrost and also to allow the migration of caribou.

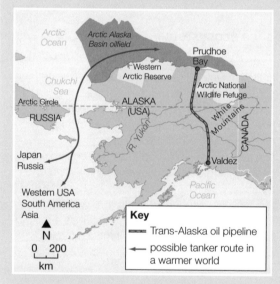

⬆ **Figure 1** *Oil exploitation in Alaska; the pipeline transports oil hundreds of kilometres south to the port of Valdez*

Over to you

For **both** Alaska and the Alps, make a list of key management strategies to reduce the impact of people on cold environments.

Student Book
pages 204–7

You need to know:

- the location and key characteristics of Svalbard
- the opportunities and challenges for development
- the human responses to these challenges.

Where is Svalbard?

Svalbard is a Norwegian territory located in the Arctic Ocean some 1000 km north of Norway (Figure **1**).

- It is the most northerly permanently inhabited archipelago in the world.
- Much of Svalbard has a dry, steppe-like climate.
- Temperatures rise to 5–6 °C in summer (July–August) and fall to −16 °C in winter (February); severe winters can see temperatures of −40 °C.
- About 60% is ice covered, 30% is barren rock (scree, moraines, fluvial deposits) and 10% is vegetated.
- *Permafrost* is almost everywhere, to a depth of 100–150 m in the valleys and up to 400 m in the mountains.
- Almost all of Svalbard's population (of about 2700) lives in the town of Longyearbyen on Spitzbergen.

▲ **Figure 1** Location of Svalbard

Challenges and opportunities for development?

Coal mining

Coal mining in Svalbard presents many challenges due to the extremely cold working conditions, the long hours of winter darkness, sea conditions affecting transportation, and the remoteness of the mines.

Originally, the centre of the coal mining industry was the Longyear Valley – a mining settlement was established at Longyearbyen in 1906. Today, only Mine 7 remains operational in the Longyear Valley. It feeds the power station in Longyearbyen, which supplies all of Svalbard's energy (Figure **2**).

The bulk of the present-day coal mining takes place at Sveagruva, 50 km south-east of Longyearbyen. The state-owned mining company, Store Norske, employs around a third of all workers on Svalbard, and extracts high-quality coal from reserves that are expected to last until about 2030.

▲ **Figure 2** The coal-fired power station at Longyearbyen is inefficient and environmentally controversial

Fishing

The cold waters of the Barents Sea to the south of Svalbard are some of the richest fishing grounds in the world, with an estimated 150 species of fish, including cod, herring and haddock. However, storms and sea ice make fishing here extremely challenging and hazardous.

Continued over ▶▶▶

Case Study

Case Study

Polar research

Svalbard has a long history of polar research, including studies of marine ecosystems, geology and meteorology.

- Russia and Poland all run permanent research stations on Svalbard.
- Currently, a great deal of research is focused on the atmosphere, analysing changes that might be linked to climate change.
- Amongst this research is the likely impact of the forecasted increases in Arctic temperature on ecosystems and physical systems, such as glaciers.

Close to Longyearbyen is the SVALSAT receiving station. Because of its latitude, a continuous download of information from 14 orbiting satellites can be obtained. NASA, NOAA and the ESA have all invested heavily in this facility.

Tourism

The opening of the airport at Longyearbyen in 1975, enabled tourism to grow significantly. Most people visit Svalbard to explore the natural environment – the glaciers, **fjords** and the wildlife (polar bears, seals, walrus) – or to study the historical development of the islands. Adventure tourism is becoming increasingly popular, with opportunities for hiking, kayaking and snowmobile safaris (Figure **3**).

Currently, some 300 local people benefit directly from being employed in the tourist sector. Tourism on land is limited to Longyearbyen and its immediate vicinity, due to the high cost of road construction and maintenance.

 Figure 3 *Sea kayaking at Spitsbergen*

Human responses to living and working in Svalbard

People can respond to the challenges of living and working in extremely cold conditions by demonstrating **resilience**, **mitigation** and **adaptation**.

Resilience

The long hours of darkness, extremely low temperatures, strong winds and, at times, heavy snowfall, require considerable resilience, particularly for those working in the coal mines or conducting research in remote environments.

Mitigation

People need to mitigate the hostile and potentially dangerous environmental conditions by wearing appropriate clothing and footwear. Houses are well insulated to protect against the cold.

Adaptation

Providing health and social care is an issue in such a remote location. Although there is a hospital and dental surgery, people with long-term needs or requiring extensive hospital attention are encouraged to fly to Norway for treatment.

Svalbard's own coal-fired power station in Longyearbyen provides energy security, vital in the cold winters. Insulated, above ground pipes carry electricity and water. Underground pipes would run the risk of rupturing during the summer as the permafrost thaws, or freezing during the winter.

Sixty second summary

- Svalbard (Norway) is the most northern permanently settled territory; it experiences very cold conditions and strong winds.
- It is mainly snow and ice or bare rock; only 10% is vegetated.
- Economic activities present significant challenges due to the harsh physical and climatic conditions.
- Coal mining, fishing and tourism are important economic activities supporting local people.
- There is a history of polar research bringing investment and employment.
- Human responses include resilience, mitigation and adaptation to the harsh environment.

 Over to you

In a list or table, summarise, the key challenges and opportunities associated with human activity in Svalbard.

5 Hazards

Your exam

(AL) *Hazards* is an **optional topic**. You must answer **one** question in Section C of Paper 1: Physical geography, from **either** *Hazards* **or** *Ecosystems under stress*. Paper 1 carries 120 marks and makes up 40% of your A Level. Section C carries 48 marks.

(AS) *Hazards* is an **optional topic**. You must answer **one** question in Section B of Paper 1: Physical geography and people and the environment, from **either** *Hazards* or *Contemporary urban environments*. Paper 1 makes up 50% of your AS Level. Section B carries 40 marks.

Specification subject content
(Specification reference in brackets)

Either *tick these boxes as a record of your revision,* **or** *use them to identify your strengths and weaknesses*

Your revision checklist

Section in student book and revision guide	☹	☺	☺	Key terms you need to understand Complete the **key terms** (not just the words in bold) as your revision progresses. 5.1 has been started for you.
The concept of hazard in a geographical context *(3.1.5.1)*				
5.1 Hazards in a geographical context				*hazards, natural hazards,*
Plate tectonics *(3.1.5.2)*				
5.2 The structure of the Earth				
5.3 Plate tectonics theory				
5.4 Plate margins and magma plumes				
Volcanic hazards *(3.1.5.3)*				
5.5 Distribution and prediction of volcanic activity				
5.6 Impacts of volcanic activity				
5.7 Human responses to a volcanic eruption				
5.8 Mount Etna, Sicily				

Seismic hazards *(3.1.5.4)*				
5.9 Earthquakes and tsunamis				
5.10 Distribution and prediction of earthquakes				
5.11 Impacts of seismic activity				
5.12 Human responses to seismic hazards				
5.13 Tōhoku – a multi-hazard environment *(your multi-hazardous environment case study)*				
Storm hazards *(3.1.5.5)*				
5.14 The nature of storm hazards				
5.15 Impacts of storm hazards				
Fires in nature *(3.1.5.6)*				
5.16 Fires in nature				
Case studies *(3.1.5.7)*				
5.17 The Alberta wildfire, 2016				
5.18 Storm hazards (Hurricane Sandy, USA, 2012 and Cyclone Winston, Fiji, 2016)				

Student Book
pages 212–15

You need to know:

- about impacts and perception of, and responses to hazards in a geographical context.

Impacts and perceptions of hazards

Hazards in a geographical context are events that threaten life. They:

- are mainly natural events (e.g. earthquakes, storms and floods)
- can be events caused by human actions (e.g. nuclear incidents)
- can be natural events as a consequence of human actions (e.g. wildfires caused by carelessness)

They become *disasters* if they cause high levels of death, injury, damage or disruption.

The impacts of hazards depend on a number of factors, such as their:

- location relative to areas of population
- magnitude and extent.

Primary impacts have an immediate effect (e.g. destruction of buildings and infrastructure). Secondary impacts happen after the event (e.g. disease in emergency camps).

How we perceive hazards is determined by culture, our experience of them and the effect that they may have on our lives. Whilst many people have no choice but to accept the risk, others often underestimate it.

Furthermore, population expansion is increasing the risks. For example:

- urbanisation – densely populated urban areas concentrate those at risk
- poverty – expense of housing leads to building on risky ground
- agriculture – harnessing the high productivity of alluvial floodplain soils puts farmers at risk.

 Big idea

Hazards pose a risk – threatening lives, possessions and the built environment. They can become disasters when extreme.

▲ **Figure 1** *Eyjafjallajökull, Iceland, 2010 – hazard or disaster?*

Human responses

The natural human response to a hazard is to reduce risk to life and equity. This involves:

- saving possessions and safeguarding property at a local level
- coordinating rescue and humanitarian aid at a global scale.

But responses can vary according to economic circumstances. *Fatalistic* attitudes (acceptance of the 'inevitable'), often found in LDEs, are in marked contrast to *adaptive* peoples in HDEs where it is more likely that

governments invest in research, prediction, preparedness and **mitigation**.

Finally, response times have been reduced by the development of the Automatic Disaster Analysis and Mapping system (ADAM), a database that pools information from the US Geological Survey, World Bank and World Food Programme. This allows almost immediate access to such information as:

- the scale of the disaster
- what supplies are available locally
- established local infrastructure.

Continued over ▶▶▶

Prediction, preparedness and mitigation

As technology increases, the methods of predicting some, but not all hazardous events improve (e.g. remote sensing and seismic monitoring). Warnings can be communicated promptly.

The hazard management cycle

For areas at risk, this illustrates both pre- and post-event situations:

- *Preparedness* – large-scale events can rarely be prevented, but education and raising public awareness can minimise impact.
- *Response* – the speed of response will depend on the effectiveness of *contingency planning*. Immediate responses focus on saving lives and coordinating medical assistance.
- *Recovery* – restoring the affected area to something approaching normality. Short-term restoration of services prior to longer-term planning and reconstruction.
- *Mitigation* – reducing the severity and impact of an event. For example:
  - earthquake and hurricane-proof buildings
  - flood barriers
  - 'defensible space' firebreaks
  - insurance.

The Park model of human response to hazards

This describes three phases following a hazard event – relief, rehabilitation and reconstruction:

▲ *Figure 2* The hazard management cycle

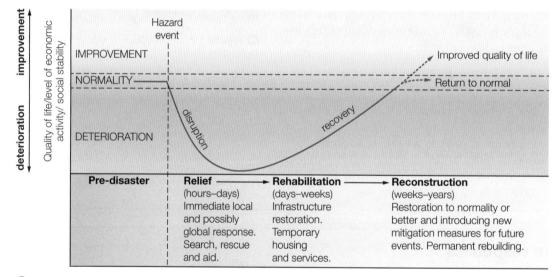

▲ *Figure 3* The Park model describes relief, rehabilitation and reconstruction after a hazard event

Sixty second summary

- Hazards are perceived according to culture, experience and the effect they may have on our lives.
- Population expansion is both increasing the risks of hazards, and decisions made that they're worth taking.
- Human responses to hazards vary from a 'do nothing' approach (fatalism) to attempts to predict, adapt to and even mitigate their impact.
- Human responses to hazards describes three phases that follow a hazard event: relief, rehabilitation and reconstruction.

Over to you

Sketch and justify Park model graphs that might apply to a:
(a) high-magnitude earthquake in an LDE built-up area
(b) landslide following prolonged rain in an HDE rural area
(c) wildfire during drought conditions close to an HDE major built-up area.

You need to know:

- about the Earth's structure and its internal energy sources.

Student Book
pages 216–17

Earth structure and internal energy sources

Edmond Halley's first theory on Earth structure (1692) suggested that the planet was made up of hollow spheres – rather like Russian nesting dolls – and that they were actually habitable! But Halley did not enjoy today's scientific expertise or images from space.

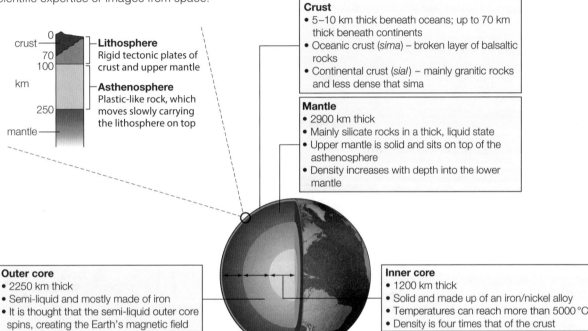

Crust
- 5–10 km thick beneath oceans; up to 70 km thick beneath continents
- Oceanic crust (*sima*) – broken layer of balsaltic rocks
- Continental crust (*sial*) – mainly granitic rocks and less dense that sima

Mantle
- 2900 km thick
- Mainly silicate rocks in a thick, liquid state
- Upper mantle is solid and sits on top of the asthenosphere
- Density increases with depth into the lower mantle

Outer core
- 2250 km thick
- Semi-liquid and mostly made of iron
- It is thought that the semi-liquid outer core spins, creating the Earth's magnetic field

Inner core
- 1200 km thick
- Solid and made up of an iron/nickel alloy
- Temperatures can reach more than 5000 °C
- Density is four times that of the crust

Lithosphere
Rigid tectonic plates of crust and upper mantle

Asthenosphere
Plastic-like rock, which moves slowly carrying the lithosphere on top

▲ **Figure 1** *The internal structure of the Earth*

Some of the Earth's internal energy (heat) may be primeval – retained by the ball of dust and gas from which the Earth evolved. But the greatest source of heat energy within the Earth is derived directly from natural radioactive decay of uranium and other elements within the core.

The phenomenal heat at the core generates convection currents within the mantle above. These currents spread very slowly within the asthenosphere and are important, but not solely responsible for the movement of tectonic plates.

 Sixty second summary

- Continental and oceanic crusts differ in depth, chemical composition and density.
- The lithosphere comprises of rigid tectonic plates of both crust and upper mantle.
- The asthenosphere beneath the upper mantle is plastic-like and can move slowly, carrying the lithosphere on top.
- As the Earth rotates, the semi-liquid outer core spins, creating its magnetic field.
- The greatest source of heat energy is derived directly from natural radioactive decay of uranium and other elements within the core.

 Over to you

Think about how you might explain the age, structure and internal energy of the Earth to an eight-year-old child. The analogies you adopt should cement your understanding.

Student Book
pages 218–19

You need to know:

- about plate tectonic theory of crustal evolution
- how tectonic plates move.

Plate tectonics theory and sea-floor spreading

Since the late 1960s, plate tectonics theory has explained the development of the Earth's surface. Submarine mapping of mid-ocean ridges and deep ocean trenches supported Alfred Wegener's controversial 1912 theory of continental drift.

Harry Hess subsequently found evidence of *sea-floor spreading*, confirmed by *palaeomagnetism* – the swapping of magnetic north and south poles every 400 000 years or so (Figure **1**).

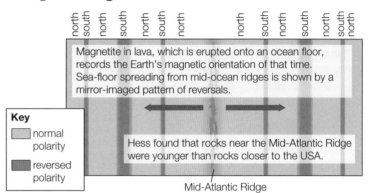

Magnetite in lava, which is erupted onto an ocean floor, records the Earth's magnetic orientation of that time. Sea-floor spreading from mid-ocean ridges is shown by a mirror-imaged pattern of reversals.

Hess found that rocks near the Mid-Atlantic Ridge were younger than rocks closer to the USA.

Key
- normal polarity
- reversed polarity

Mid-Atlantic Ridge

Figure 1 *Sea-floor spreading*

Tectonic plates and plate movement

The Earth's surface is made up of irregularly shaped (*lithospheric*) tectonic plates floating on the plastic *asthenosphere* underneath.

- Oceanic plates of dense basaltic rocks (*sima*) are continually being formed at mid-ocean ridges and destroyed at deep ocean trenches. Continental plates of less dense granitic *sial* are permanent and much older.
- Oceanic and continental plates move between 2 cm to 16 cm a year.
- Radioactive decay within the Earth's core generates huge temperatures. *Hot spots* around the core create rising thermal currents, before cooling and sinking again.
- Conventional explanations suggest mantle convection drives tectonic plates. But **gravitational sliding** related to lithospheric thickening may be more significant – 'pushing' the older part of the plate in front (**ridge push**) and, following *subduction*, pulling the rest of the plate with it (**slab pull**).

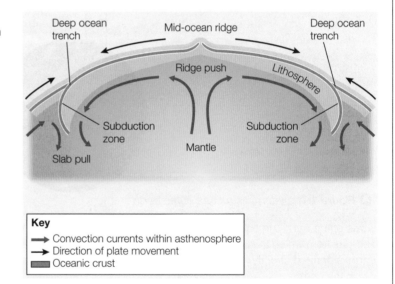

Key
- → Convection currents within asthenosphere
- → Direction of plate movement
- ▬ Oceanic crust

 Figure 2 *Factors explaining tectonic plate movement*

Sixty second summary

- The movement of tectonic plates results in processes that destroy oceanic plate at subduction zones and create new (oceanic) plate material along mid-ocean ridges.
- Wegener's theory of continental drift was supported some 50 years later by Hess's discovery of sea-floor spreading.
- Palaeomagnetism reveals symmetrical patterns of magnetic reversals moving away from mid-ocean ridges.
- Gravitational sliding is now thought to be a more significant mechanism for plate movements than mantle convection.

Over to you

Make sure you understand and learn the key terms of this topic (identified in *italics* and **bold**).

You need to know:

- about the different kinds of plate margins and magma plumes
- landforms associated with plate tectonics.

Student Book
pages 220–3

Plate margins

Constructive (divergent) plate margins

Fluid, basaltic magma forces its way to the surface, causing shallow-focus earthquakes and creating broad, flat shield volcanoes in rift valleys. **Submarine volcanoes** along mid-ocean ridges sometimes rise above sea level to become volcanic islands.

There are two types of divergence:

- *Continental* (e.g. East Africa's Great Rift Valley) – stretching and collapsing of the continental crust to create massive *rift valleys* separated by blocks of land (*horsts*).
- *Oceanic* (e.g. the Mid-Atlantic Ridge) – sea-floor spreading on either side of mid-ocean ridges rising up to 4000 m above the ocean floor. Sections widen at different rates due to **transform faults** (Figure 1).

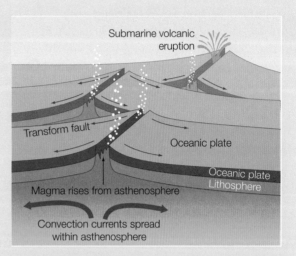

Figure 1 *Transform faults along a mid-ocean ridge*

Destructive (convergent) plate margins

There are three types of convergence:

- *Oceanic meets continental* (e.g. the Pacific coast of South America) – dense oceanic plate is **subducted** under the folding continental plate and destroyed by melting in the *Benioff Zone*. This creates rising plumes of acidic magma which erupts explosively from steep-sided, composite, andesitic and rhyolitic volcanoes. Friction in the Benioff Zone leads to intermediate- and deep-focus earthquakes (Figure **2**).
- *Oceanic meets oceanic* (e.g the Mariana Trench in the western Pacific) – the faster or denser plate subducts leading to earthquakes and the formation of a deep-ocean trench. Benioff Zone melting results in rising **magma plumes** and submarine volcanoes which may create *island arcs* (Figure **3**).
- *Continental meets continental* (e.g. the Himalayas) – less dense continental plates, uplift and buckle into high fold mountains. There is no subduction, so no magma to form volcanoes. But powerful, shallow-focus earthquakes can be triggered.

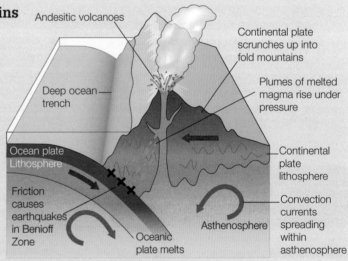

Figure 2 *Oceanic plate meets continental plate*

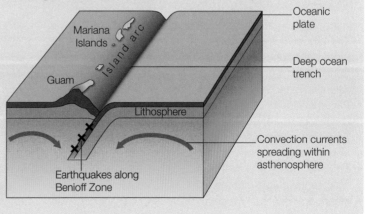

Figure 3
Oceanic plate meets oceanic plate

Continued over ▶▶▶

Conservative plate margins

Two plates move past each other in different directions (e.g. Haiti) and/or at different rates (Figure **4**). Friction between the plates build stresses and trigger powerful, shallow-focus earthquakes when they slip. There are no volcanoes because there is no magma.

This type of earthquake occurs along California's San Andreas fault system (Figure **5**).

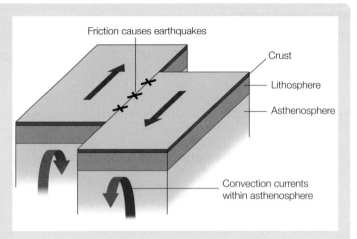

Figure 4 ○
Conservative plate margin

Magma plumes

Around 95% of the world's earthquakes and most volcanoes are located along plate margins.

Even seemingly notable exceptions, such as the volcanic Hawaiian Islands, are explained in part by plate movements.

- Concentrated *hot spots* associated with radioactive decay within the Earth's core heat the lower mantle creating localised thermal currents where *magma plumes* rise vertically.
- The plumes then 'burn' through the lithosphere to create volcanic activity on the surface.
- Hot spots remain stationary, so movement of the overlying plate results in the formation of a chain of active and extinct shield volcanoes (e.g. the Hawaiian Islands, near the centre of the Pacific Plate).

Plate margins and processes are summarised on page 223 in the student book.

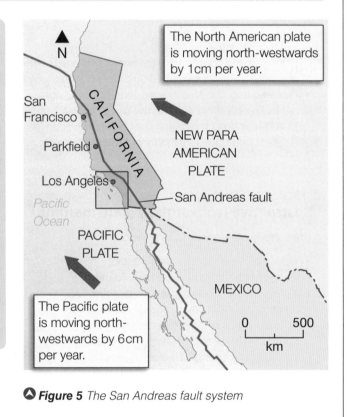

The North American plate is moving north-westwards by 1cm per year.

NEW PARA AMERICAN PLATE

San Andreas fault

The Pacific plate is moving north-westwards by 6cm per year.

 Figure 5 *The San Andreas fault system*

Sixty second summary

- Plates separate at constructive margins causing shallow-focus earthquakes and basaltic volcanic eruptions.
- The processes and resulting landforms at constructive (divergent) plate boundaries differ according to whether the plate margin is in a continental (e.g. Great African Rift Valley) or oceanic area (e.g. Mid-Atlantic Ridge).
- Plates collide at destructive plate margins causing intermediate- and deep-focus earthquakes, and explosive volcanic eruptions.
- Subduction only takes place where a denser, oceanic plate meets another plate.
- Fold mountains and shallow-focus earthquakes are created where two continental plates meet (e.g. the Himalayas).
- Hot spots heat the lower mantle creating localised thermal currents where magma plumes rise vertically. If these plumes occur within a plate, they may burn through the lithosphere creating shield volcanoes.

Over to you

List and explain all the key terms of this topic (identified in *italics* and **bold**).

Student Book
pages 224–7

You need to know:
- about the spatial distribution of volcanoes
- their magnitude, frequency, regularity and predictability.

Spatial distribution of volcanoes

Figure **1** shows the relationship between earthquakes, volcanoes and plate margins.

- Volcanic activity (**vulcanicity**) is evident at constructive and destructive plate margins.
- Volcanoes are absent at conservative margins.
- Some volcanoes occur within plates (e.g. the Hawaiian hot spot), and along rift valleys (e.g. the Great African Rift Valley).
- Location affects the type and magnitude of eruption, and the type of magma.

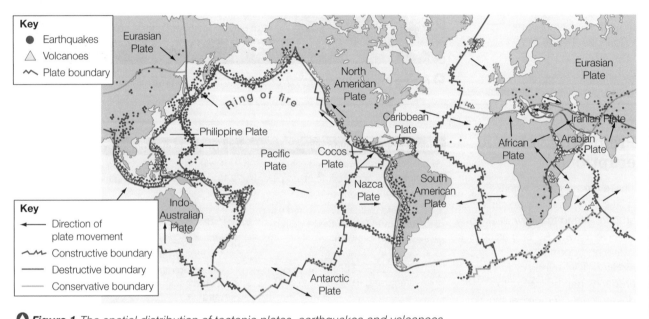

Figure 1 *The spatial distribution of tectonic plates, earthquakes and volcanoes*

Magnitude

Since 1982 the magnitude of volcanic eruptions has been measured using the *Volcano Explosivity Index (VEI)* (typical Hawaiian eruptions are 1, Eyjafjallajökull was a 4).

When, rather than if, a **supervolcano** erupts again, the VEI might exceed 7 and change global climates for years afterwards!

Volcanic classifications are more useful if related to tectonics by including details of the type of magma and even frequency (Figure **2**).

Figure 2
A classification of volcanic eruptions

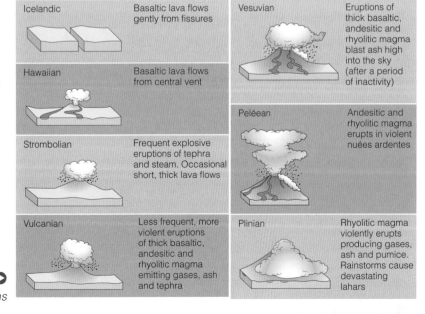

Continued over >>>

Frequency and regularity

Frequency and regularity of eruptions are rarely, if ever, predictable to any degree of accuracy. The uncomfortable reality is that volcanoes such as Chances Peak, Montserrat (see 5.7) have proved to be dormant rather than extinct. Even for those volcanoes known to be dormant, relying on average cycles of activity can only alert volcanologists of the necessity for heightened observation. For example, Mount Vesuvius, Italy, is one of the world's most carefully monitored volcanoes (Figure **3**).

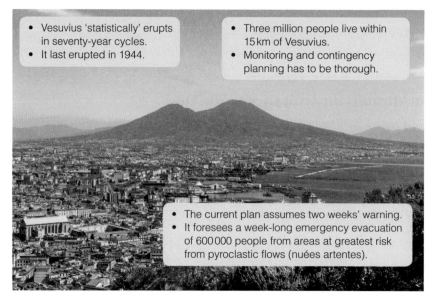

- Vesuvius 'statistically' erupts in seventy-year cycles.
- It last erupted in 1944.

- Three million people live within 15 km of Vesuvius.
- Monitoring and contingency planning has to be thorough.

- The current plan assumes two weeks' warning.
- It foresees a week-long emergency evacuation of 600 000 people from areas at greatest risk from pyroclastic flows (nuées artentes).

Figure 3 *Mount Vesuvius, overlooking the Bay of Naples*

Predicting volcanic eruptions

Unlike earthquakes, where there is little or no warning, volcanic eruptions tend to follow weeks of seismic activity. As long as both active and dormant volcanoes are monitored, warnings of imminent eruptions can be issued to governments and civil authorities (Figure **4**).

Prediction is generally very successful and can be aided by evidence from previous eruptions – such as **lahar** and pyroclastic deposits following river valleys. As part of the contingency planning, hazard maps can be produced to identify those areas most at risk, which are therefore prioritised for evacuation to safe zones.

What is being monitored?	What does it indicate?
Seismic activity is measured using seismometers and recorded using a seismograph.	Microquakes indicate rising magma fracturing and cracking the overlying rocks.
Ground deformation is measured using tiltometers and laser-based electronic distance measurement.	Bulging (inflation) of the ground is caused by rising magma. Slope angles and increasing distance between points can be accurately measured.
Upward movement of iron-rich magma is measured using magnetometers.	Changing magnetism within the volcano is a common geophysical indication of rising magma.
Rising groundwater temperature and/or gas content is measured using hydrological instrumentation.	Rising magma will both heat groundwater and corrupt it with gases such as sulphur (increasing its acidity).
Warning signs such as small eruptions, emissions of gases, landslides and rockfalls can be recorded in real time using remote sensing equipment.	Remote solar-powered digital camera surveillance records physical changes in and around the main crater. Thermal imaging and sampling of (poisonous) emissions such as chlorine can also be included in remote sensing.

Figure 4 *Methods and instruments used in the monitoring of active and dormant volcanoes*

 Sixty second summary

- Volcanoes are found at constructive and destructive plate margins and, sometimes, within plates (e.g. hot spots).
- Volcanoes may be classified in terms of the magnitude, characteristics/ frequency of eruptions, or by type of magma.
- Scientific techniques can be used to monitor activity and predict when an eruption will occur. Eruptions tend to follow weeks of seismic activity.
- Evidence from previous eruptions, such as lahars and pyroclastic deposits, can be used to inform the extent of evacuation plans.

 Over to you

Refresh your knowledge of remote sensing methods used to monitor volcanic activity. Then outline their purpose and value.

You need to know:
- what factors affect the impacts of volcanic activity.

Student Book
pages 228–9

The impacts of volcanic activity

The impacts of volcanic activity depend upon the type of volcano.

- *Fissure eruptions* of basic lava create extensive, featureless lava plateaus (e.g. the Deccan Traps in central India). They represent the largest contributors to global climate change and large-scale landscaping.
- *Basic shield volcanoes* are vast, shallow-sided and broad (e.g. Hawaii and Iceland). Their eruptions are gentle enough to become tourist attractions.
- *Acid dome volcanoes* are steep-sided convex cones. Explosive eruptions of pyroclastic flows have deadly impact (e.g. Mount Pelée, Martinique).
- *Composite cones (strato-volcanoes)* are formed from alternating eruptions of ash, **tephra** and lava which build up in layers (Figure **1**).
- *Calderas* are the remains of volcanoes that have collapsed into their own emptied magma chambers following violent eruptions. Lakes fill the vast pit craters or they are flooded by the sea (e.g. Santorini in the Mediterranean).

 Big idea

Volcanoes are many things. They are ruthless destroyers of lives and property. But they are also benefactors. Think about why?

▲ *Figure 1 The classic composite cone of Mount Fuji, Japan*

Two contrasting eruptions

Both Eyjafjallajökull and Mount Merapi were Peléean/Plinian eruptions (VEI 4), yet both had contrasting social, economic and environmental impacts.

Eyjafjallajökull, Iceland (2010)	Mount Merapi, Java, Indonesia (2010)
• Local, national, regional and global impacts. • Subglacial eruption – considerable flooding. • No fatalities or injuries. • Vast ash cloud that spread over a wide area. • 100 000 flights cancelled and over 10 million people left stranded – the cost to airlines was US$1.7 billion.	• Local and national impacts. • Pyroclastic flows, lahars, ash falls and lava bombs. • 350 fatalities caused by fires, burns, respiratory failure and blast injuries – notably those refusing evacuation and/or returning to their homes between eruptions. • 350 000 displaced. • Considerable disruption to national aviation.

 Sixty second summary

- The impacts of volcanic activity depend upon the type of volcano – viscous lavas are more likely to be explosive.
- Volcanic activity can have social, economic and environmental impacts.
- Vulnerability, level of preparedness, wealth and land use affect the risk of a disaster.
- The 2010 eruptions of Eyjafjallajökull and Mount Merapi illustrate how the impacts of volcanic activity can differ dramatically.

 Over to you

Create a Venn diagram to compare and contrast the 2010 eruptions of Eyjafjallajökull and Mount Merapi.

You need to know:

* about short- and long-term responses to a volcanic eruption, Montserrat.

Student Book
pages 230–1

Montserrat, 1995 – present

Montserrat's Chances Peak was thought to be extinct, but ash eruptions in July 1995 prompted an immediate response of scientific monitoring and subsequent evacuations (Figure 1).

On 25 June 1997, Chances Peak catastrophically erupted, engulfing the south of the island in pyroclastic flows. Further eruptions and associated evacuations occurred in December 2006, July 2008 and February 2010 (Figure 2).

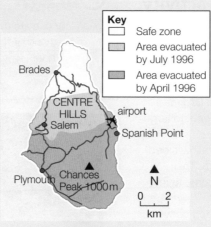

Key
- ☐ Safe zone
- ☐ Area evacuated by July 1996
- ☐ Area evacuated by April 1996

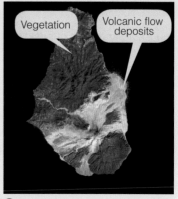
Vegetation · Volcanic flow deposits

Figure 1 Montserrat in the Lesser Antilles island arc

Figure 2 Satellite image after the 2010 eruption

Short-term responses	Long-term responses
Montserrat Volcano Observatory set up in 1995, successfully predicted the 1997 eruption.	UK aid funded three-year redevelopment programme.
Southern exclusion zones delimited.	In 1998, the Montserrat population were granted full UK residency rights. This resulted in a top-heavy population structure as many younger people did not see an economic future on the island.
NGOs (e.g. Red Cross) provided temporary schooling, medical support and food.	By 2005 many people had returned, but the population had halved.
Warning systems set up (e.g. sirens and via the media).	Cash-cropping re-established as ash, lava and lahar deposits broke down. But two-thirds of the island remained an exclusion zone.
US troops and the British Navy helped with the evacuation.	Rebuilding adventure tourism with the volcano itself as an attraction – a new airport, hotel and dive shop built.
£17 million in UK aid paid for temporary buildings and water purification systems.	UK financial aid since 1995 has exceeded £420 million.

Figure 3 Summary of short- and long-term responses

Sixty second summary

* Unexpected eruptive activity of Chances Peak in 1995 prompted monitoring and subsequent evacuation of the south of the island.
* A catastrophic eruption in 1997 engulfed the region with pyroclastic flows.
* This, and subsequent eruptions in 2006, 2008 and 2010, prompted evacuation of much of the population, resulting in a top-heavy population structure.
* By 2005 many people had returned, but large areas of the south remained uninhabited.
* Farming of the fertile, volcanic soils, and development of adventure tourism, have the potential to rebuild Montserrat's economy.

Over to you

Practise describing **four** immediate and **four** long-term responses to the Montserrat eruptions.

You need to know:

* about impacts and human responses to a recent volcanic event, Mount Etna in Sicily.

Student Book
pages 232–3

Mount Etna, Sicily, Italy

Mount Etna is Europe's most active volcano. It is a continually monitored strato-volcano with a complex, dynamic geology of four summit craters, and numerous secondary (parasitic) cones and fissures. Spectacular activity, involving a wide variety of eruptive styles, is frequent.

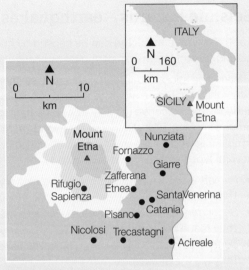

Figure 2 *Mount Etna eruption, December 2015*

Figure 1 *Location of Mount Etna, Sicily, Italy*

Date	Nature of eruption	Impacts and human responses
March 1983	Fissure effusion	Slow-moving lava redirected using huge earthworks
Dec 1991–Mar 1993	Effusive lava flows and Hawaiian fountaining	Zafferana Etnea threatened and defended by: • an earth dam • blasted rock and concrete blocks • diverting lava flow into a new man-made channel
Jul–Aug 2001	Seven flank fissures erupt effusively	• Significant damage to tourist facilities • Nicolosi threatened • Ash falls close Catania airport
Oct 2002	Strombolian, Hawaiian fountaining and *phreatomagmatic* flank eruptions	• The most explosive flank eruption in 150 years • Lava flows threatened Rifugio Sapienza • Catania airport closed again
Nov 2009	Summit Strombolian eruptions	4.4 magnitude earthquake
Apr 2010	Summit ash eruption	Eruption increased crater width
Jan–Oct 2011	• Summit Hawaiian fountaining and lava flows. • Ash column of several km high	Lava flow descending western slope successfully diverted from the ski resort at Sapienza Refuge
Oct 2013	Renewed summit eruptions	
Jan–Jun 2014	Summit Strombolian eruptions	• Lava flows travelled east and north-east • Catania Airport temporarily closed
Dec 2015	Summit Strombolian eruptions: • lava fountaining reached 1 km high • ash column rose 7 km.	• Sulphur dioxide plumes drift to North Africa, Iran and Turkmenistan • Short closure of Catania Airport

Figure 3 *Summary of Mount Etna eruptions 1983–2015*

Sixty second summary

* Mount Etna is Europe's most active volcano.
* A strato-volcano, with summit craters, numerous secondary cones and fissures – resulting in varied eruptive behaviour.
* Continual remote sensing to keep tourists and the local community safe.
* To protect property, lava flows have been successfully diverted.

Over to you

Summarise this page in no more than **75** words.

You need to know:

- the nature of seismic hazards
- the underlying causes of earthquakes and tsunamis.

Student Book
pages 234–5

Seismic hazards – earthquakes and tsunamis

Earth shaking (**seismicity**) can be caused by human activities (e.g. mining), but around 95% of the world's earthquakes occur along or near tectonic plate margins.

Friction and sticking between plates causes enormous pressures and stresses. When they fracture, a series of seismic shockwaves are sent from the *focus* to the surface. The *epicentre* (directly above the fracture) experiences the most intense ground shaking, with deterioration beyond. Tremors usually last for less than a minute followed by several weeks of aftershocks as the crust settles. There are several types of seismic wave (Figure **1**):

- *Primary or pressure (P)* waves are like high frequency sound waves. They reach the surface fastest.
- *Secondary or shear (S)* waves shake like a skipping rope. They reach the surface next.
- *Surface Love (L) and Rayleigh (R)* waves cause the most damage. They are the slowest.

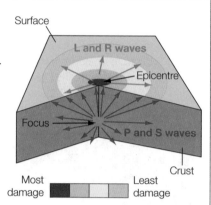

△ **Figure 1** *The focus (hypocentre) and epicentre of an earthquake*

Tsunamis

Tsunamis are a highly destructive series of waves characterised by:

- very low wave height, but very long wavelength
- very high speed – between 640–960 km per hour
- a long time between each wave (the wave period) – between 10 and 60 minutes
- on approaching the coast, the waves slow and pile up as a massive wall of water.

Effective warning systems give many hours warning. Without this, the first sign – the apparent draining away of the sea in front of the tsunami (known as a drawdown) – will be too late.

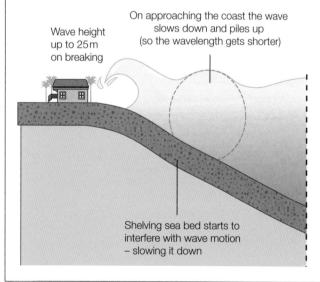

▽ **Figure 2** *Characteristics of a tsunami*

- Most earthquakes are caused by friction along plate margins, creating stresses in the lithosphere that fracture and send shockwaves to the surface.
- Seismic shockwaves take different forms, travel at different speeds and cause varying amounts of damage.
- Tsunamis are usually generated by seismic activity which displace huge volumes of water creating destructive waves.

 Over to you

Make sure that you can explain earthquakes and tsunamis clearly, using appropriate terminology.

Student Book
pages 236–7

You need to know:

- about the spatial distribution of earthquakes
- their magnitude, frequency, regularity and predictability.

Spatial distribution of earthquakes

The relationship between earthquakes, volcanoes and tectonic plate margins was established in 5.5:

- Earthquake activity is evident at constructive, destructive and conservative plate margins.
- The number, intensity, depth and wider extent of earthquakes vary according to the type of margin (Figure **1**).

Magnitude, frequency, regularity and predictability

Earthquake *magnitude* is measured on the logarithmic Richter scale. Each number is ten times the magnitude of the one before it, so a slight increase in value equates to an enormous effect on the ground.

Earthquake *damage* is measured on the Modified Mercalli Intensity scale, using observations of the earthquake's impact. The scale ranges from I (imperceptible) to XII (catastrophic).

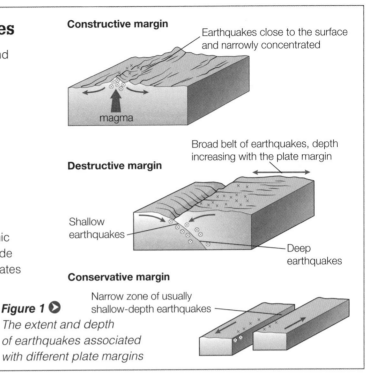

Constructive margin
Earthquakes close to the surface and narrowly concentrated
magma

Destructive margin
Broad belt of earthquakes, depth increasing with the plate margin
Shallow earthquakes
Deep earthquakes

Conservative margin
Narrow zone of usually shallow-depth earthquakes

Figure 1 ❯
The extent and depth of earthquakes associated with different plate margins

Mitigation – the best defence?

Earthquakes cannot be accurately predicted, although clues can occur before they strike (e.g. microquakes, ground bulging). Risk assessment, contingency planning and earthquake engineering is required. For example:

- earthquake-resistant construction
- Geographic Information Systems (GIS) used to prepare hazard maps and to inform the planning of urban development
- public education, such as earthquake preparation checklists and practising evacuation drills.

Rolling weights on roof to counteract shock waves

Panels of marble and glass flexibly anchored to steel superstructure

Reinforced lift shafts with tensioned cables

'Birdcage' interlocking steel frame

Reinforced latticework foundations deep in bedrock

Rubber shock absorbers between foundations and superstructure

▲ **Figure 2** *A modern earthquake-resistant building*

Sixty second summary

- The number, intensity, depth and wider extent of earthquakes varies between plate margins.
- Earthquake magnitude is measured on the logarithmic Richter scale.
- Earthquake damage is measured on the Modified Mercalli Intensity scale.
- Earthquakes cannot be accurately predicted.
- Mitigation may be the best defense, including the construction of earthquake-resistant buildings and the use of GIS to prepare hazard maps to inform plans for urban development.

Over to you

Should scientists have to strike a balance between attempting to predict earthquakes, and earthquake mitigation? How should that balance be weighted, and why?

You need to know:

- about the primary and secondary impacts of seismic activity
- about the Indian Ocean tsunami in 2004.

Student Book
pages 238–41

The impacts of seismic activity

Impacts of an earthquake depend not only on its magnitude, depth and surface geology, but also on population density, the design of buildings, and the time of day.

Primary effects – immediate impact	Secondary effects – impacts resulting as a direct consequence	Long-term impacts
Ground shaking damages/destroys buildings and infrastructure	• Fires caused by broken gas pipes and power lines • Hindered emergency services • Diseases spread from contaminated water	• Higher unemployment • Repair and reconstruction may take months or years • Longer-term illness and/or reduced life-expectancy
Schools, colleges and universities destroyed	• Education suspended for immediate future	• A 'lost generation' to develop economy in future
Immediate deaths and injuries	• Bodies that are not buried spread diseases (e.g. cholera) • Injuries not receiving prompt treatment	• Trauma and grief may last months or years • Disability and reduced life-expectancy
Shocked, hungry people forced to sleep outside	• NGOs provide tents, water and food	• Emergency pre-fabricated homes may become permanent fixtures
Landslides and avalanches	• Further deaths and injuries • Flooding from blocked rivers	• Reduced agricultural productivity • Permanent disruption to drainage patterns
Liquefaction of saturated soils	• More building collapses, deaths and injuries	• Repairs/reconstruction difficult and expensive.
Damage to power stations	• Power cuts restrict emergency services/ medical care	• Repairs/reconstruction very expensive
Panic, fear and hunger	• Civil disorder and looting • Police/army intervention	• Problem restoring trust in neighbours and civil authorities

⬆ **Figure 1** *Impacts of earthquakes*

The Indian Ocean tsunami, 26 December 2004

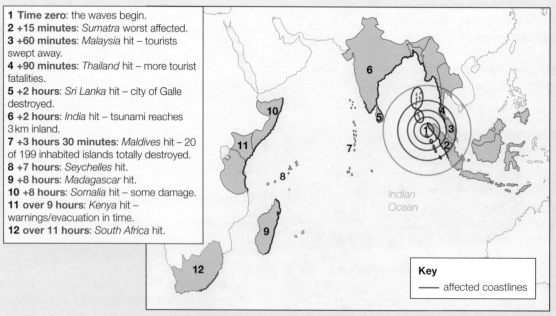

1 Time zero: the waves begin.
2 +15 minutes: *Sumatra* worst affected.
3 +60 minutes: *Malaysia* hit – tourists swept away.
4 +90 minutes: *Thailand* hit – more tourist fatalities.
5 +2 hours: *Sri Lanka* hit – city of Galle destroyed.
6 +2 hours: *India* hit – tsunami reaches 3 km inland.
7 +3 hours 30 minutes: *Maldives* hit – 20 of 199 inhabited islands totally destroyed.
8 +7 hours: *Seychelles* hit.
9 +8 hours: *Madagascar* hit.
10 +8 hours: *Somalia* hit – some damage.
11 over 9 hours: *Kenya* hit – warnings/evacuation in time.
12 over 11 hours: *South Africa* hit.

Indian Ocean

Key
—— affected coastlines

⬆ **Figure 2** *Countries affected by the 2004 Indian Ocean tsunami*

The tsunami was caused by a magnitude 9.0–9.3 earthquake (the second largest in history), which ruptured 1000 km of sea bed off the coast of Sumatra. Countries around the entire span of the Indian Ocean were affected in what proved to be one of the world's worst natural disasters (Figure **2**).

Primary and secondary impacts

Primary impacts

Up to 300 000 fatalities

Infrastructure destroyed

Coastal settlements devastated (e.g. the city of Banda Aceh in Sumatra

Vegetation and top soil stripped up to 800 m inland

Secondary impacts

Widespread homelessness (e.g. 500 000 forced into refugee camps in the worst hit region of Aceh Province, Indonesia)

Economies devastated (e.g. fishing, agriculture, tourism)

Water supplies and soils contaminated by salt water

Negative multiplier effects weaken economies (e.g. loss of deep sea trawlers resulted in reduced catches, which were less likely to reach market because of fewer refrigerated trucks and damage to infrastructure)

⬆ *Figure 3 Impacts of the powerful waves along the southern Indian coastline*

Immediate responses

- Massive international relief efforts involved more than 160 aid organisations and UN agencies.
- Foreign military troops provided assistance.

Long-term responses

- Programmes of reconstruction implemented (but many still in tents one year on).
- Political barriers delayed aid distribution (e.g. in Sri Lankan areas held by rebel Tamil Tigers).
- Existing government prejudices highlighted (e.g. the Indian *Dalits* 'underclass' were ignored).
- Tourist resorts quickly rebuilt (e.g. Phuket, Thailand).
- Tsunami warning system, including specific contingency planning.

Six

Sixty second summary

- Impacts depend on magnitude, depth and surface geology. Also on population density, building design and the time of day.
- Earthquakes can have environmental, social, economic and political impacts on an area and its population.
- The 2004 Indian Ocean tsunami was one of the world's worst natural disasters.
- Effects of the tsunami included nearly 300 000 fatalities, widespread destruction and homelessness, contamination of water supplies and economic devastation.
- The international aid response was huge, but varied in its effectiveness.

Over to you

Make sure that you understand the tectonics responsible for this event. Review 5.3, 5.4, 5.9 and 5.10 and note the specific plate names and type of margin in order to inform your examination answers.

You need to know:

- about short- and long-term responses to seismic hazards
- the effects of and responses to the Port-au-Prince earthquake, Haiti, 2010

Student Book
pages 242–5

Short- and long-term responses to seismic hazards

As with volcanic eruptions, each new seismic event studied teaches us more, and so preparedness, short-term (immediate) and long-term responses can continue to improve.

Big idea

Seismic hazards tend to mirror all natural hazards in that 'those with least suffer most'.

The Port-au-Prince earthquake, Haiti, 2010

Haiti is a multi-hazard LDE vulnerable to tropical storms, flooding, landslides, periodic droughts and earthquakes. It is also the poorest country in the western hemisphere, blighted by political instability, corruption, poor infrastructure, social inequality, exclusion and unrest.

A complex strike-slip fault, following the conservative plate margin running through Haiti and the Dominican Republic, had been jammed since the last earthquake of 1751. On 12 January 2010 the stress of the surrounding rocks was finally overcome and the plates were released, resulting in a catastrophic, shallow-focus, magnitude 7 earthquake (Figure 1).

Primary impacts

- Over 230 000 died in less than 60 seconds.
- 50% of Port-au-Prince buildings collapsed (including key government buildings).
- Over 180 000 homes were damaged and 1.5 million people made homeless.
- Infrastructure and nearly 5000 schools were damaged or destroyed.

Secondary impacts

- Strong aftershocks were recorded.
- Government was crippled.
- There was general lawlessness.
- Cholera killed over 1500 and 1.5 million people were still homeless one year later.

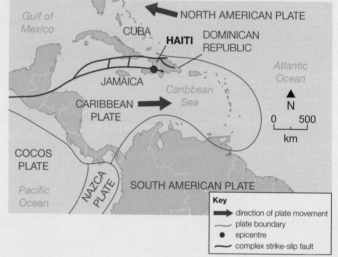

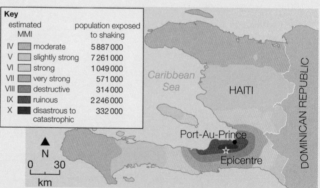

Figure 1 *Caribbean plate margins and intensity map of Haiti using the Modified Mercalli Intensity (MMI) scale*

Figure 2 *High-density, poorly built concrete buildings destroyed in the capital, Port-au-Prince*

Short-term and long-term responses

Short-term (immediate) responses	Long-term responses
Rescue efforts – international search teams and locals employed by the UNDP (United Nations Development Project) pulled survivors out from the debris and cleared roads.	*Aid* – the Haiti Relief Fund manages an $11.5 billion reconstruction package with controls in place to prevent corruption. Reconstruction is due to be completed by 2020.
Infrastructure – in Port-au-Prince, the US military took control of the airport to speed up the distribution of aid.	*Economy* – UN strategy was developed to create new jobs in clothing manufacture, tourism and agriculture, and also to reduce the effects of uncontrolled urbanisation.
Security – 16 000 UN troops and police restored law and order.	
Food – the UN World Food Programme provided basic food necessities.	*Food* – farming sector was reformed to encourage greater self-sufficiency and less reliance on food imports.
Water and sanitation – the UK Disaster Emergency Committee (DEC) provided bottled water and purification tablets, and built 3000 latrines.	*Water and sanitation* – supplies and services eventually restored.
Health – emergency surgeries established to perform life-saving operations.	*Health* – new focus on follow-up care, including mental health care.
Shelter – Up to 1.5 million people occupied over 1100 camps. Emergency shelter provided for 1.9 million people.	*Buildings* – hospitals, schools and government buildings rebuilt to new life-safe building codes. Local people employed as construction workers. New settlements built away from high risk areas (e.g. unstable hillsides).

 Figure 3 *Short-term and long-term responses to the 2010 Haiti earthquake*

Tents and tarpaulins provided emergency shelter for 1.9 million people.

By the first anniversary of the earthquake, 1.5 million people were still homeless.

Reconstruction allowed improvement:
- slums were demolished
- new, life-safe homes are more affordable, safe and sustainable.

 **Figure 4** *Hopeless or hopeful? – tent city, Haiti 2010*

Sixty second summary

- Preparedness is the best defence for mitigating against the impact of seismic hazards.
- Haiti is a multi-hazard LDE vulnerable to tropical storms, flooding, landslides, periodic droughts and earthquakes.
- The 2010 Haiti earthquake was a shallow-focus, magnitude 7.0 event resulting in devastating primary and secondary impacts.
- The international response to the disaster included the UN, World Food Programme, the UK's Disaster Emergency Committee and the US military.

Over to you

Hazard **impacts** are often characterised as environmental, social and economic. Why would a Venn diagram be ideal for summarising these **responses** under the same environmental, social and economic headings?

Case Study

You need to know:

- about the Tōhoku earthquake and tsunami in Japan, 2011
- the effects and responses.

The Tōhoku earthquake and tsunami, Japan, 2011

On average Japan records 1500 earthquakes every year – around one-third of the world's total!

At 2.46 p.m. on Friday 11 March, a magnitude 9.0 earthquake occurred under the Pacific Ocean (Figure **1**).

- A 400–500 km segment of the North American Plate, snagged by the subducting Pacific Plate, suddenly released upwards by between 5 and 10 m, triggering a tsunami.
- Japan's tsunami warning system kicked in, but people along a 3000 km stretch of coastline had just 30 minutes to escape.
- Ten waves, each about 1 km apart, slowed and piled up as they reached the shallower coastal water, overwhelming tsunami defence walls and surging up to 10 km inland.

The earthquake was the biggest ever recorded in Japan, and had global significance. It:

- moved the entire island of Honshu 2.4 m closer to North America
- shifted the Earth's axis by at least 10 cm
- made Earth days shorter by 1.8 microseconds
- calved 125 km² of icebergs from the Antarctic coast
- caused visible waves in Norwegian fiords.

Primary effects

- Ground shaking caused buildings to collapse, fractured gas pipes and started fires.
- The tsunami swept inland, devastating nearly 500 km².
- In Tokyo, skyscrapers 'started shaking like trees', but their earthquake-resistant design meant damage was limited.
- Over 18 000 were dead or missing, although Japan's tsunami warning system saved many lives.

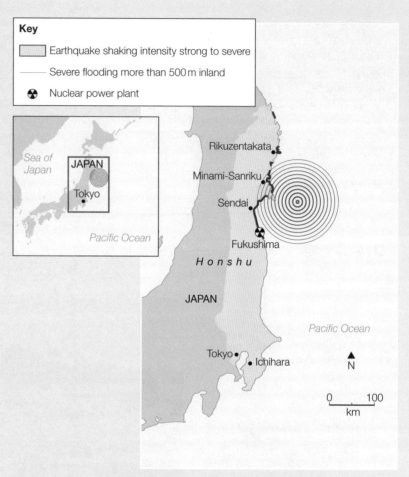

Key

- ▨ Earthquake shaking intensity strong to severe
- ─ Severe flooding more than 500 m inland
- ☢ Nuclear power plant

☝ **Figure 1** *The earthquake epicentre and area affected*

☝ **Figure 2** *Primary effects in the town of Minami-Sanriku (see Figure 1)*

Secondary effects

- Half a million people were homeless. For weeks, 150 000 people lived in temporary shelters.
- At least 1 million homes were left without running water and almost 6 million homes lost their electricity supply.
- There were shortages of food, water, petrol and medical supplies.
- In the two weeks after the earthquake, there were more than 700 aftershocks, causing concern and further damage.
- Explosions and radiation leaks at the Fukushima Daiichi nuclear power plant in the days after the earthquake spread fear around the world and caused panic selling across global stock markets.

Immediate responses

- Mobilisation of 100 000 soldiers to organise rescue work and distribute blankets, bottled water, food and petrol.
- Global aid poured in from, for example, the UK, USA and China.
- An exclusion zone was set up around the Fukushima Daiichi nuclear plant (Figure 3); homes were evacuated and iodine tablets distributed.
- There were no reports of looting or violence.

Long-term responses

- Japan coped well with the earthquake, but tsunami defences were inadequate, prompting a review of future contingency planning.
- In 2013 Japan unveiled a more informative tsunami warning system.
- To stimulate the economy, Special Zones for Reconstruction were designated with relaxed planning regulations and tax incentives.
- Japan's immediate electricity shortfall was met through imports of oil and gas. Only in 2016 were nuclear reactors issued licenses to restart.
- By the end of 2017 most of the evacuation orders of the Fukushima prefecture had been withdrawn, but 55 000 people still lived as evacuees.
- There are fears of an increase in the incidence of juvenile thyroid cancer. (Figure 4)
- Estimated at US$300 billion, this is the most costly natural disaster in history – a repair bill to be met by more government borrowing.

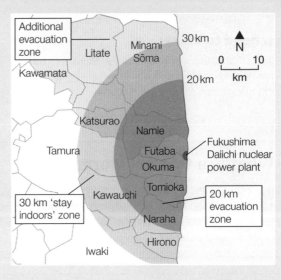

△ **Figure 3** *Fukushima Daiichi 'evacuation' and 'stay indoors' orders were repeatedly revised*

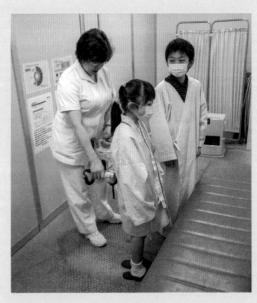

△ **Figure 4** *200 000 children now suffer from precancerous thyroid abnormalities*

Sixty second summary

- In March 2011, a magnitude 9.0 earthquake created a series of tsunami waves that hit the Japanese coast 30 minutes later.
- Japan's tsunami warning system saved lives, but defence walls were overwhelmed causing widespread destruction.
- At an estimated US$300 billion, this was the most costly natural disaster in history.
- Explosions at the Fukushima Daiichi nuclear power plant caused global concern – exclusion zones were set up and the majority of Japanese reactors shut down.
- Immediate and long-term responses confirmed Japan's global reputation as disciplined, and culturally, economically and technologically sophisticated.

Over to you

List the bullet points from the Sixty Second Summary. Underneath **each** point add supporting evidence and/or examples.

Student Book
pages 250–5

You need to know:

- the characteristics and spatial distribution of tropical storms
- the hazards associated with tropical storms
- the magnitude, frequency, regularity and predictability of tropical storms.

What is a tropical storm?

A tropical storm (cyclone in India, hurricane in the North Atlantic, typhoon in south-east Asia) can cause extensive damage and loss of life in many parts of the tropics. A tropical storm has average wind speeds in excess of 120 km/h (75 mph) and can be 500 km in diameter.

Formation and development of a tropical storm

Once a tropical storm has started to form, it will soon develop its distinct and clearly defined rotation:

- Warm, moist air rises rapidly in its centre, creating a central vortex, to be replaced by air drawn in at the surface.
- The eye is often characterised by a column of dry, sinking air. The eye wall is the most damaging part of a storm.
- The rising air cools, condenses and towering cumulonimbus clouds form.
- When condensation occurs, latent heat is released, which effectively powers the storm.
- Cloud and rain extend in a series of waves.
- A storm starts to decay as it reaches land, as the supply of energy and moisture is cut off.

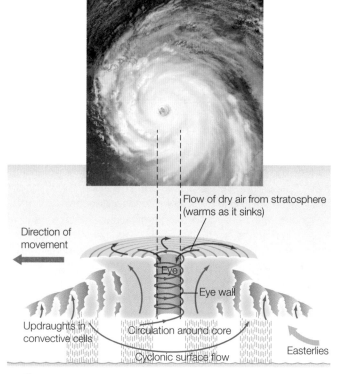

▲ **Figure 1** *Cross-section through a tropical storm: note the degree of symmetry around the eye*

Distribution of tropical storms

Several factors affect the distribution of tropical storms:

- *Oceans* – tropical storms derive their moisture from the oceans and peter out on reaching land.
- *High temperatures* – a sea-surface temperature in excess of 26 °C.
- *Atmospheric instability* – most likely to form in regions where warm air is being forced to rise, such as the ITCZ.
- *Rotation of the Earth* – 'spin' is needed to initiate the rotation of a tropical storm. Storms do not usually form between 5 °N and 5 °S.
- *Uniform wind direction at all levels* – winds from different directions at altitude 'shear off' the vertical development of a tropical storm, restricting height and intensity.

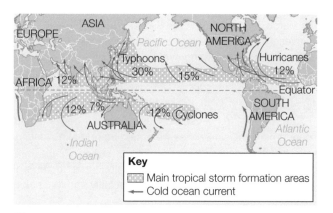

▲ **Figure 2** *Tropical storms are mainly found in the tropics. The percentages show the proportion of the total global number of storms in each region*

What are the hazards associated with tropical storms?

Strong winds

- Wind with speeds in excess of 120km/h (75mph) are capable of causing significant damage to structures, infrastructure and communication networks.
- Damaged power lines often lead to disruption through electricity cuts and, occasionally, fires.

Landslides

- Many landslides are triggered by tropical storms.
- Intense rainfall increases soil pore water pressure, weakens cohesion and triggers slope failure.
- The additional weight of water exacerbates the problem.
- In 1998 Hurricane Mitch triggered multiple landslides that killed 18000 people in Central America.

Storm surges

- A storm surge is caused by the intense low atmospheric pressure of the tropical storm (forcing the sea to rise vertically) together with the powerful surface winds.
- They are a major cause of widespread devastation and loss of life.
- They also inundate agricultural land with saltwater and debris, pollute freshwater supplies and destroy housing and infrastructure.

Coastal and river flooding

- A tropical storm can generate torrential rainfall in just a few hours.
- This can trigger flash flooding at the coast, particularly in urban areas where the drainage system cannot cope.
- A tropical storm weakens inland, but river flooding may still occur due to rain intensity.

Frequency and magnitude

Figure **3** shows there is no clear evidence that the numbers nor intensity (magnitude) of storms are increasing as global temperatures increase.

Predictability

To some extent, tropical storms can be predicted – they are mostly restricted to the tropics and do not usually occur close to the Equator. They also mostly occur from late summer into autumn with a peak from August through to October.

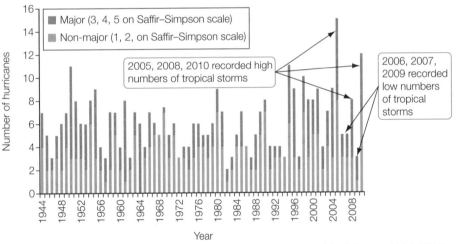

- Major (3, 4, 5 on Saffir–Simpson scale)
- Non-major (1, 2, on Saffir–Simpson scale)

2005, 2008, 2010 recorded high numbers of tropical storms

2006, 2007, 2009 recorded low numbers of tropical storms

Figure 3 *Number of hurricanes in the North Atlantic/Caribbean, 1944–2010*

See page 255 of the student book for regularity of tropical storms.

The Saffir–Simpson scale is explained on page 253 of the student book.

Sixty second summary

- Tropical storms are immensely powerful and can cause extensive damage and loss of life.
- Several factors are needed for a tropical storm to form, primarily high ocean temperatures, atmospheric instability, 'spin' resulting from the Earth's rotation and uniform wind directions at altitude.
- Associated hazards include strong winds, storm surges, flooding and landslides.
- There is little evidence that climate change affects frequency and magnitude.
- There is some degree of regularity and predictability, although their tracks can be erratic.

Over to you

Practise drawing a simplified labelled sketch of Figure **1**. Add further annotations to summarise the factors affecting the formation of tropical storms and the hazards resulting from them.

You need to know:
- the impacts of tropical storms
- how the impacts of tropical storms can be reduced.

Student Book
pages 256–61

Impacts of tropical storms

In common with all natural hazards, there are a range of impacts of tropical storms (Figure **1**).

 Big idea

The impact of a natural hazard may be reduced by preparedness, mitigation, prevention and adaptation.

Impact	Definition	Example
Primary	Initial and direct impacts of a tropical storm – strong winds, storm surge, heavy rain, flooding.	Hurricane Sandy, USA (2012) caused extensive destruction to Eastern USA.
Secondary	A consequence of the primary impacts, such as inland river flooding and landslides.	Landslides caused by Hurricane Mitch in Central America (1998) killed thousands.
Environmental	Impacts on the environment – seawater contamination of freshwater habitats and aquifers, destruction of coastal environments and pollution.	Cyclone Nargis, Myanmar (2008) caused devastation to crops by saltwater and pollution.
Social	Impacts on people, including death, injury and disruption to people's everyday lives.	Hurricane Katrina, USA (2005) displaced over one million people from New Orleans.
Economic	The financial costs of a tropical storm to local people and governments. Increasingly, the financial burden is supported by the insurance industry.	Hurricane Katrina (2005) is the USA's costliest tropical storm disaster, with damages estimated at US$150 bn.
Political	Tropical storms can lead to political issues of command and control. To what extent is it a local, regional, national or even an international issue? A disaster of national emergency status often receives greater personnel and financial support.	The Myanmar government did not encourage international support in the wake of Cyclone Nargis. This left many local communities having to cope themselves. A total of 80 000 people died.

⬆ **Figure 1** *Different types of impacts of tropical storms*

Reducing the impacts of tropical storms

Although it is not possible to prevent a tropical storm from forming or making landfall, it is possible to take measures aimed at reducing its impacts.

These measures can be behavioural (e.g. increasing people's preparedness) or structural, which can involve small-scale building adaptations, as well as larger-scale constructions, such as sea walls. In locations that are very prone to tropical storms (e.g. the Philippines, Bangladesh), people have had to adapt by using structures such as cyclone shelters.

There are several approaches that may be adopted to reduce the impact of a natural hazard – preparedness, mitigation, prevention and adaptation.

See page 261 of the student book for a case study in Bangladesh.

Preparedness

Increasing people's awareness and through their actions minimise the likely impact of the hazard:

- through education and public awareness campaigns, people can make minor structural improvements to buildings (e.g. stronger doors and windows)
- by preparing emergency supplies and plan evacuation routes
- by insuring property
- by prediction of the likely course using satellite/ radar tracking and computer models such as SLOSH.

Mitigation

Actions aimed at reducing the severity of an event and lessening its impact. This can include structural responses, disaster aid and insurance cover.

- *Structural responses* offer some protection from storm surges by soft-engineering (planting trees and building up beaches) or hard-engineering schemes, such as constructing sea walls. Grants are available to make homes more **resilient**.

- *Disaster aid* can come from trading blocs (e.g. the EU) or from international bodies (e.g. the World Bank or the United Nations). Charities and other NGOs also provide valuable support, often reflecting generous donations from members of the public. Aid can take two forms:
  - o immediate humanitarian relief in the form of search and rescue, food, water, medicine and shelter
  - o longer-term reconstructional aid that seeks to support recovery and reconstruction.

- *Insurance cover* is widely used to mitigate the effects of tropical storms, particularly in HDEs. However, the rich can afford insurance, the poorest in society cannot. Many of those who were most affected by Hurricane Katrina in New Orleans 2005 were the poor who did not have insurance; they refused to be evacuated in order to safeguard their property.

See 'Protecting Galveston' panel on page 259 for an example of a structural response.

Prevention

Prevention is actions aimed at reducing the chance of large-scale events from starting (e.g. wildfires).

Scientists have unsuccessfully tried cloud seeding (dropping crystals into clouds to cause rain) to dissipate tropical storms. Now the focus is on forecasting, together with mitigation and adaptation, in reducing the impacts of tropical storms.

Adaptation

Tropical storms cannot be prevented so people simply have to accept that natural events are inevitable and learn to live with the threat but do what they can to minimise the risks.

Land-use zoning aims to reduce the vulnerability of people and property at the coast. Most commonly this allows only low-value land uses (e.g. recreation) to occupy the coastal strip. In parts of north-eastern Florida, coastal properties are raised above the ground on stilts and have non-residential functions on the ground floor (Figure **2**).

 Figure 2 *Adaptation to coastal flooding in Florida, USA. The residential parts of the structures are raised above ground, with garages on the ground floor*

Sixty second summary

- There are several significant impacts of tropical storms – environmental, social, economic and political.
- There are several approaches to reducing the hazards associated with tropical storms – preparedness, mitigation, prevention and adaptation.
- Preparedness involves increasing people's awareness and ability to respond appropriately when warnings are issued.
- Mitigation can involve structural responses (such as a sea wall) or behavioural responses (such as disaster aid and insurance).
- Prevention is not really feasible given the nature of tropical storms.
- Adaptation involves learning to live with the threat and being able to respond accordingly, (e.g. cyclone shelters).

Over to you

Produce a spider diagram to clarify the various impacts of tropical cyclones.

Student Book
pages 262–7

You need to know:
- the characteristics, formation and impacts of wildfires
- the strategies for managing wildfires.

What is a wildfire?

Wildfire is the generic name used for an uncontrolled rural fire (*bushfires* in Australia, *brushfires* in North America). They affect different layers of vegetation (Figure **1**).

With the exception of Antarctica, every continent experiences conditions favourable for the ignition of wildfires. As populations have grown, and with more people moving into rural areas, the risk has increased.

Wildfires release carbon stored in trees, plants and peat, enhancing the greenhouse effect and increasing the likelihood of wildfires (positive feedback loop).

Wildfires are the result of certain conditions.

Vegetation type – fuel characteristics

The type and amount of fuel (vegetation) influences the intensity (the output of heat energy) and rate of spread (degree of threat). For example, grassland fires rarely produce the same intensity as forest fires, and the eucalyptus is fire-promoting – oils within the leaves can explode!

Climate and weather conditions

Most wildfires occur during or after prolonged dry periods. Strong, dry winds blowing from continental interiors or deserts help the drying process and are ideal conditions for lightning storms – a common form of ignition. Wind strength determines the rate of spread (Figure **2**).

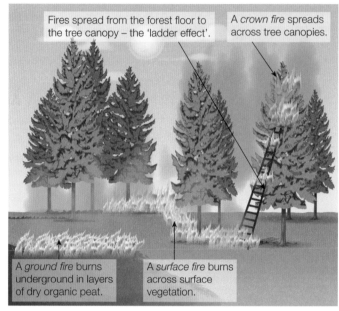

Fires spread from the forest floor to the tree canopy – the 'ladder effect'.

A *crown fire* spreads across tree canopies.

A *ground fire* burns underground in layers of dry organic peat.

A *surface fire* burns across surface vegetation.

Figure 1 *Three types of wildfire*

1. Desert winds originate from a clockwise flow of air around a high-pressure system east of the Sierra Nevada mountains.

2. Air from the mountains is compressed and warmed, becoming less humid. This lower humidity dries out vegetation and can fan any existing fires.

3. Winds squeeze through canyons with gusts between 65 and 95 km/h

4. Strong winds create turbulence and can make interstate travel difficult.

Figure 2 ▶
The Santa Ana winds play a significant role in the development of and path taken by wildfires in southern California

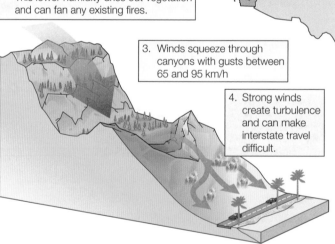

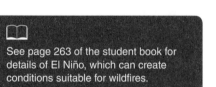
See page 263 of the student book for details of El Niño, which can create conditions suitable for wildfires.

Causes of wildfires

Most fires that threaten life and residential areas, particularly in woodland close to large urban areas such as Sydney and Los Angeles, are the result of human actions (such as discarded cigarettes and campfires).

Heat transfer processes (radiation, conduction, convection) preheat trees, forest litter and also vegetation ahead of the flames, enabling rapid spread of the fire through spot fires.

Impacts of wildfires

	Primary impacts	Secondary impacts
Environmental	• Destruction of habitats and ecosystems • Death and injury of animals, which impacts on food chains and food webs • Short-term surge of carbon dioxide due to the burning of carbon stores (trees) • Pollution from smoke and toxic ash	• Lack of vegetation depletes nutrient stores, increases leaching and risk of flooding • Increased carbon emissions impact on the greenhouse effect and so climate change • Effects on ecosystem development – secondary succession
Social	• Loss of life and injury • Displacement of people • Damage to power lines/communications	• New employment required • Behavioural adaptations – people may have to abide by new rules and regulations
Economic	• Damage/destruction of structures • Financial (loss of earnings, damage costs) • Destruction of businesses, crops, livestock	• Costs of rebuilding or possible relocation • Replacement of infrastructure, crops, livestock • Cost of future preparedness and mitigation
Political	• Actions of emergency services • Responses of government • Response of local authorities and emergency services in the immediate aftermath	• Strategies for preparedness and mitigation • Decisions about replanting forests, compensation, future regulations, etc. • Review laws/advice regarding use of countryside for leisure

🔺 **Figure 3** *Primary and secondary impacts of wildfires*

Strategies for managing wildfires

There are four strategies for managing wildfires – preparedness, mitigation, prevention and adaptation.

Preparedness

Early detection and suppression of wildfires can take the form of:

• voluntary rural firefighting teams
• warnings issued as fire risk increases
• firebreaks around properties.

Prevention

• Public awareness can prevent fires starting.
• Many countries operate 'fire bans' during times of high risk.
• Controlled burning reduces the fuel store but it may get out of control, impact on the natural ecosystems and release carbon dioxide.

Mitigation

Reducing the impact before, during and after the event.

• Early detection by cameras and drones, satellites and infrared sensors.
• *Back burning* ahead of the fire to remove the fuel. Rivers may also control the spread.
• Disaster aid and fire insurance can mitigate the effects.

Adaptation

Allowing wildfires to burn themselves out.

• Burning old/diseased wood stimulates fresh growth.
• Regulations can restrict access to areas at risk of wildfire.
• Simple buildings made of natural products which will not cause pollution if burnt.

 Sixty second summary

• There are several types of wildfire, e.g. ground, surface and crown fires.
• Wildfires have environmental, social, economic and political impacts.
• Preparedness involves increased awareness and warnings.
• Mitigation is identification and intervention to reduce impacts.
• Prevention can reduce human-induced fires.
• Adaptation involves allowing some fires to burn in fire-prone areas.

 Over to you

Create a simple diagram to describe the different types of wildfire.

Case Study

You need to know:

- the factors that caused the Alberta wildfire in Canada
- the environmental, social, economic and political impacts
- the responses to the wildfire.

Student Book
pages 268–71

The Alberta wildfire, 2016

In May 2016 a huge wildfire struck parts of Canada's Alberta province, forcing the evacuation of 90 000 residents of Fort McMurray as the fire destroyed 2400 homes and businesses.

What were the causes and contributory factors?

The fire ignited in a forested area south-west of Fort McMurray but the precise cause remains unknown. A shift in the wind direction took the blaze into the outskirts of Fort McMurray, the largest settlement in the area.

Figure **1** shows how erratic the fires were. It clearly shows 'spotting', where wind-carried burning embers ignite fires well ahead of the fire front – even across a one-kilometre river in places.

A lack of winter snowfall and an early spring snowmelt, combined with above average temperatures, provided dry conditions ideal for an outbreak. The intensity of the fire created its own weather patterns, including strong winds and lightning, which led to the ignition of additional fires in the area.

Climate scientists have linked the fire to a strong El Niño effect that may well have resulted in the unusually warm and dry early spring conditions.

△ **Figure 1** *Thermal fire map of Fort McMurray; spotting can be easily seen. Notice that the fire even bridged the Athabasca and Clearwater Rivers.*

What were the impacts of the wildfire?

Environmental	Social	Economic	Political
• The fire severely affected the forest ecosystem due to the scorched soil and burned tree roots. • There was a risk of re-ignition of the scorched peaty soils. • Toxins released from burning trees and buildings created air pollution; several million tonnes of carbon dioxide were released into the atmosphere. • Ash was washed into water courses leading to water pollution.	• 90 000 people were forced to flee Fort McMurray. • 2400 homes and other buildings were burned down in parts of Fort McMurray. • Jobs and livelihoods were affected and movement in the area was restricted. • Increased levels of anxiety about the future. • Power supplies were disrupted. • Water supplies became contaminated.	• Initial insurance company estimates suggested CAN\$9bn of damage was inflicted upon Fort McMurray. • About a third of the 25 000 workers in the nearby oil sands industry had to be evacuated from work camps. The fire is estimated to have cost the industry CAN\$1bn. • Transport in the region was seriously affected, including at the nearby international airport.	• The fire has stimulated political debate about the possible impacts of climate change. • Government liaised with emergency services in implementing evacuation programmes. • The Alberta government implemented a programme of phased and safe re-entry. • Coordination of reconstruction programmes for buildings, services and infrastructure.

△ **Figure 2** *Impacts of the Alberta wildfire, 2016*

An evacuation of the size as that of Fort McMurray created social, economic and political impacts. The effectiveness of the measures taken were proven by the fact that no one was killed or even injured.

Figure 3 shows some of the neighbourhoods that were most affected.

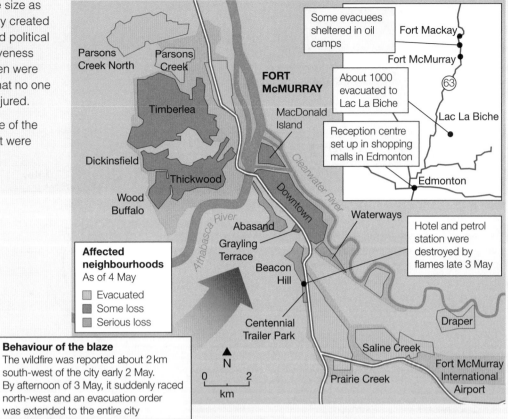

 Figure 3 *The neighbourhoods in Fort McMurray most affected by the wildfires. Compare this with the satellite image in Figure 1.*

What were the responses to the wildfire?

The initial response to the outbreak was monitoring and forecasting of the track of the fire. Subsequent responses included:

- A well-organised evacuation of Fort McMurray prevented deaths and injuries. Aircraft were used to evacuate some of the oil sands workers.
- The Alberta government declared a state of emergency triggering support from the Canadian armed forces.
- The Alberta government provided evacuees with CAN$1250 per adult and CAN$500 per dependant to help cover living expenses.
- The Canadian government pledged long-term aid to support the rebuilding process.
- A benefit concert in Edmonton, 'Fire Aid', raised money for those affected by the disaster.

Sixty second summary

- The Alberta wildfire devastated the Canadian city of Fort McMurray in May 2016.
- Very dry conditions, high temperatures and strong winds contributed to the enormity of the blaze (with possible links to an El Niño event).
- Impacts included evacuation of 90 000 people from Fort McMurray, temporary cessation of oil production from nearby oil sands and widespread forest destruction.
- Well planned and executed emergency procedures resulted in no deaths or major injuries.
- Political institutions provided relief and long-term responses.

Over to you

Produce a summary information poster or infographic outlining the causes, impacts and responses of the Alberta wildfire.

You need to know:

- the impacts of and responses to Hurricane Sandy
- the impacts of and responses to Cyclone Winston
- factors responsible for the contrasting impacts and responses of the two storms.

Student Book
pages 272–7

Hurricane Sandy, USA, 2012

Impacts

In October 2012, Hurricane Sandy began life as a tropical storm over the warm waters of the Caribbean. It passed through Jamaica, Haiti, Cuba and the Bahamas with sustained winds of 160 km/h. The storm then travelled along the US east coast and merged with a weather system from the west, becoming an 'extra-tropical cyclone', devastating large areas of north-east USA.

The torrential rainfall, ferocious winds and powerful storm surges had huge impacts:

- 233 people were killed
- thousands of homes were destroyed; 200 000 were homeless in Haiti alone
- millions of people were left without electricity
- ruptured gas pipes caused many fires
- sand and debris disrupted the road network
- power lines were destroyed; 70% of Jamaica's population were without power
- huge disruption in New York City as streets, tunnels and subway lines were flooded
- widespread disruption across a huge area of the Caribbean and 24 US states
- total cost of damage was US$75 billion.

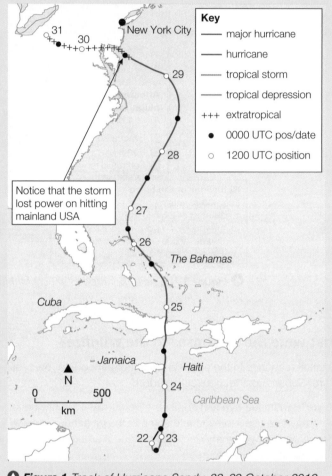

▲ **Figure 1** *Track of Hurricane Sandy, 22–29 October 2012*

Preparation

Measures were taken to reduce the impacts on people and property.

In Jamaica:

- schools and government buildings were closed
- houses were reinforced
- people stocked up on provisions
- Kingston Airport was closed.

In the USA:

- power companies were prepared to repair power lines
- the government was prepared to supply aid
- the military were put on alert
- schools were closed, hurricane centres opened and people evacuated.

Responses

Short-term responses included:

- provision of food, water and shelter
- the UN and World Food Programme sent supplies for 500 000 people in Cuba
- restoration of power in the USA
- the American Red Cross supplied over 4000 volunteers to help those affected
- In New York, government provided emergency supplies of petrol.

Long-term responses included:

- a live telethon concert raised over US$20 million
- a 'Day of Giving' raised US$17 million
- the US government approved a US$50 billion relief package.

Cyclone Winston, Fiji, 2016

Impacts

In February 2016, Fiji was struck by the southern hemisphere's strongest tropical cyclone ever recorded with sustained wind speeds of over 230 km/h. Winston started as a small tropical storm to the east of Vanuatu, then strengthened over the warm seas, before heading westwards for Fiji, increasing in intensity.

The exceptionally strong winds caused huge damage:

- 44 people were killed
- 250 000 people without clean water and sanitation
- Over 40 000 homes were damaged or destroyed
- Communications were lost for several days
- 80% of the population lost power
- Over 225 schools were damaged or destroyed
- Damage to crops caused food prices to rise
- The agricultural sector lost over US$54 million
- Total cost of damage estimated at US$1.4 billion.

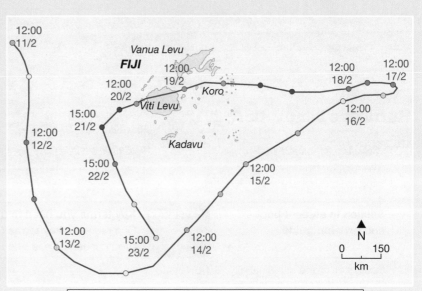

 **Figure 2** *The circuitous track of Cyclone Winston shows the difficulty in prediction. Red figures show time and date.*

Preparation

Many lives were saved by:

- monitoring and forecasting so warnings could be issued
- opening around 700 cyclone shelters
- encouraging people to leave their homes and seek shelter
- the military being put on alert to support the relief operation
- suspending public transport and advising people not to travel.

Responses

Short-term responses included:

- provision of shelter, food, water and medicine
- declaring a state of emergency
- financial assistance and supplies from the international community
- the international airport reopened after two days
- telephone services were restored
- all schools reopened after two weeks
- power supplies were restored after three weeks
- clearing waste and debris to make way for reconstruction.

Long-term responses included:

- US$9 million of governmental financial support for reconstruction
- 'Help for Homes' – a programme of house building for those unable to rebuild
- New Zealand supplied personnel and financial relief, as well as aircraft and helicopters
- Australia provided financial relief and also sent air support and emergency personnel
- charities and NGOs raised money for those affected by the disaster.

 Sixty second summary

- Hurricane Sandy killed 233 people and left hundreds of thousands homeless and without electricity.
- Cyclone Winston killed 44 people, flattened crops and destroyed 40 000 homes.
- The impacts of the two storms were similar – many people were affected and long-term economic issues associated with crop damage.
- Responses to the two storms were reasonably swift, although it took two days for Fiji's airport to reopen and allow international aid to be received.

Over to you

Create a table to contrast the impacts of and responses to the tropical storms in the USA and Fiji. Focus on the similarities and differences.

6 Ecosystems under stress

Your exam

(AL) *Ecosystems under stress* is an **optional topic**. You must answer **one** question in Section C of Paper 1: Physical geography, from **either** *Hazards* **or** *Ecosystems under stress.*
Paper 1 carries 120 marks and makes up 40% of your A Level. Section C carries 48 marks.

Your revision checklist

Specification subject content (Specification reference in brackets)

Either tick these boxes as a record of your revision, or use them to identify your strengths and weaknesses

Section in student book and revision guide	☹	😐	☺	Key terms you need to understand Complete the **key terms** (not just the words in bold) as your revision progresses. 6.1 has been started for you.
Ecosystems and sustainability (3.1.6.1)				
6.1 An introduction to biodiversity				*ecosystem, biodiversity,*
6.2 Ecosystems, development and sustainability				
Ecosystems and processes (3.1.6.2)				
6.3 The nature of ecosystems – energy flows and trophic levels				
6.4 Nutrient cycles				
6.5 Terrestrial ecosystems				
Biomes (3.1.6.3)				
6.6 Biomes				

6.7 The tropical rainforest biome			
6.8 The savanna grassland biome			

Ecosystems in the British Isles over time *(3.1.6.4)*

6.9 Ecosystems and vegetation succession in the British Isles			

Marine ecosystems *(3.1.6.5)*

6.10 Marine ecosystems			

Local ecosystems *(3.1.6.6)*

6.11 Urban wasteland ecosystems			

Case studies (3.1.6.7)

6.12 Regional ecological change – Exmoor, UK			
6.13 Wicken Fen, Cambridgeshire, UK			

Student Book
pages 282–5

You need to know:

- the definition of biodiversity and how it can be measured
- causes of biodiversity decline
- recent trends in biodiversity.

Big idea

Biodiversity is crucial to the functioning of ecosystems and can be considered the foundation to all life on Earth.

What is biodiversity?

Biodiversity is defined by the Convention on Biological Diversity (1992), as 'the variability among living organisms from all sources including terrestrial, marine and other aquatic **ecosystems** and the ecological complexes of which they are a part; this includes diversity within species, between species and of ecosystems.'

Measuring biodiversity

Several indicators can be used to monitor trends in biodiversity:

- *Species richness* – essentially this is the number of different species.
- *Population number* – the number of genetically distinct populations of a particular species.
- *Genetic diversity* – the variation in the amount of genetic information within and among individuals of a population, a species, an assemblage or a community.
- *Species evenness* – measurement of how evenly individuals are distributed among species.

Trends in biodiversity

Many animal and plant populations have declined in numbers, geographical spread, or both. Human activity has increased the extinction rate by at least 100 times compared to the natural rate.

The WWF's Living Planet Index (LPI) is a widely accepted measure of the state of the world's biological diversity. It is based on population trends of vertebrate species from terrestrial, freshwater and marine habitats.

During the period 1970–2000, the index fell by some 40% (Figure **1**). Those areas experiencing the most rapid decline in biodiversity include the Amazon basin, the Great Lakes region of Eastern Africa, the Indus valley and parts of the Middle East.

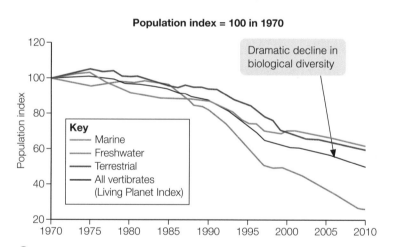

Population index = 100 in 1970

Dramatic decline in biological diversity

Key
- Marine
- Freshwater
- Terrestrial
- All vertibrates (Living Planet Index)

⬣ **Figure 1** *This graph shows a decline in all species, according to the WWF's Living Planet Index*

Recent trends

In recent years the decline has started to show signs of slowing down. This is due to:

- The designation of protected areas, now covering nearly 13% of the world's land area.
- Increasing recognition of indigenous and local community-managed areas.
- Adoption of policies and actions for managing invasive alien species – about 55% of countries have legislation to prevent the introduction of new alien species and control existing invasive species.

- Regulations supporting sustainable harvesting, reduced pollution and habitat restoration.
- International financing for biodiversity conservation is estimated to have grown by about 38% in real terms since 1992 and now stands at over US$3 billion per year.

Read about the plight of bees on page 283 of the student book.

Causes of biodiversity decline

Much biodiversity decline is due to deforestation, the expansion of intensive farming and urbanisation (Figure **2**).

In marine ecosystems, inevitably it is the waters closest to major populations, particularly in parts of Asia, where species face the greatest threats. For example:

- Some 35% of mangroves have been lost in the last two decades as coastal fringes have been developed for tourism or intensive fish farming.
- Roughly 20% of the world's coral reefs have been destroyed by overfishing, pollution and, more recently, coral bleaching associated with climate change (Figure **3**).

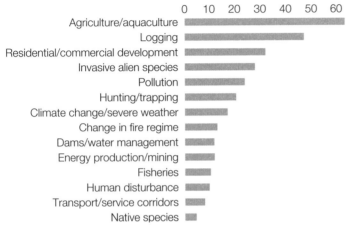

Proportion of threatened species affected, %

◆ **Figure 2** *Causes of biodiversity decline. More land was converted for agricultural use between 1950 and 1980 than between 1700 and 1850*

◆ **Figure 3** *Coral bleaching*

Potential impacts of declining biodiversity

Reduction in biodiversity can affect stores and flows such as decomposition rates, vegetation biomass production and, in the marine environment, fish stocks. Scientists are concerned that at some point in the future a threshold may be crossed when positive feedbacks lead to a catastrophe equal to that of global warming.

Biodiversity loss has negative effects on several aspects of human well-being, such as vulnerability to natural disasters, energy security, food security and access to clean water and raw materials.

In order to emphasise the importance of ecosystems, attempts are being made to assign a monetary value that takes into account the benefits for recreation, water supply and cultural significance (Figure **4** puts a monetary value on coral reefs). This may help to further slow down, or even reverse, biodiversity decline.

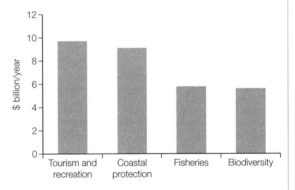

◆ **Figure 4** *Global annual value of coral reefs. Knowing the monetary value of ecosystems may make financial institutions take more of an interest*

 Sixty second summary

- Biodiversity refers to the number and types of organisms in an ecosystem or environment.
- In recent decades, many regions have recorded a decline in biodiversity, particularly as a result of deforestation, agricultural developments and urbanisation.
- Biodiversity decline can impact on natural disasters, resource insecurity and ill health.
- Responses include the creation of protected land and the increasing use of regulations to prevent harmful practices.

Over to you

From memory, define the **two** key terms in this spread (biodiversity and ecosystem), and then check them against the glossary and/or student book. If necessary, refine your definitions and write them out again.

You need to know:

- the connections between ecosystem services and human well-being
- the role of ecosystems in population growth and economic development.

Student Book
pages 286–7

Why are ecosystems important for human development?

Figure **1** shows the connections between ecosystem services (the functions offered by ecosystems) and human well-being (effectively human development).

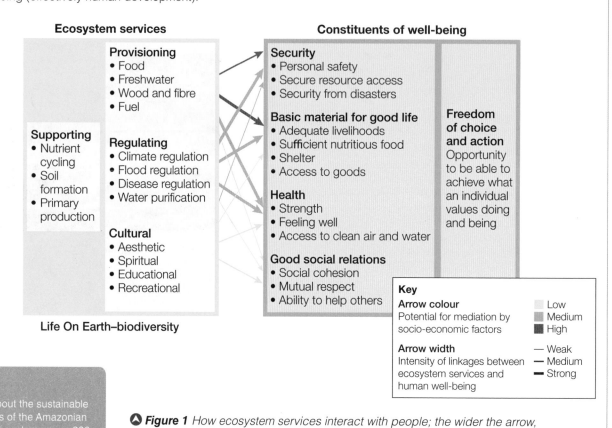

Read about the sustainable practices of the Amazonian Achuar people on page 286 of the student book.

🔺 *Figure 1 How ecosystem services interact with people; the wider the arrow, the more intense the link*

Ecosystems, population growth and economic development

The last 500 years have seen an explosion in population growth with the global population reaching 7.4 billion in 2016. It is projected to reach 10 billion by 2050.

- As population increases, agricultural productivity has to increase to avoid starvation and famine.
- Increased productivity requires increased intensification (using chemicals and mechanisation) and the conversion of marginal 'natural' land into productive farmland.
- Inevitably, population growth, together with increased economic pressures such as resource exploitation, will lead to ecosystem degradation – pollution, deforestation, desertification and biodiversity reduction.

Sixty second summary

- Ecosystem services (functions offered by ecosystems) are clearly linked to human activity (security, health, materials) and to well-being.
- Population growth and economic development exploit ecosystem services and risk causing harmful impacts of selective ecosystems, leading to ecosystem degradation (e.g. soil erosion, draining wetlands).
- Rich and diverse rainforest ecosystems have been destroyed at an alarming rate to address the needs of an ever-expanding global population.

Over to you

Identify a number of clear links between ecosystem services and human well-being.

Student Book
pages 288–93

You need to know:

- that ecosystems are natural systems with inputs, outputs, stores and flows (transfers)
- that ecosystems include a variety of plants, animals and reducer organisms at each trophic level
- that multiple food chains interconnect to create complex food webs.

The structure of ecosystems

Ecosystems at any scale (from local to global) strive to achieve a balance between living organisms in the *biotic* environment (plants and animals, including dead and decomposing matter) and inorganic non-living substances in the *abiotic* environment (minerals, water and gases in the soil, relief, drainage and climatic variables). These 'ecological systems' have:

- inputs (e.g. solar energy)
- outputs (e.g. nutrient leaching)
- stores (e.g. nutrients held in the biomass)
- flows (e.g. infiltration)
- feedback loops (e.g. germinating seeds increase numbers of plants and therefore biomass – a positive feedback).

See page 289 in the student book for the application of systems concepts.

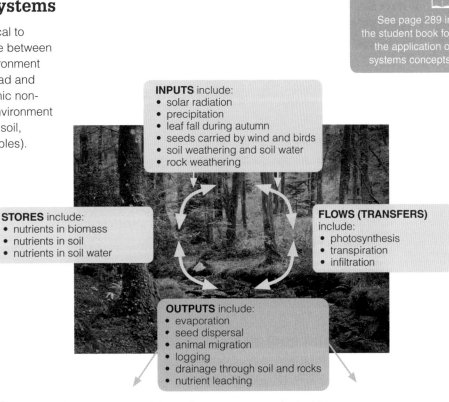

INPUTS include:
- solar radiation
- precipitation
- leaf fall during autumn
- seeds carried by wind and birds
- soil weathering and soil water
- rock weathering

STORES include:
- nutrients in biomass
- nutrients in soil
- nutrients in soil water

FLOWS (TRANSFERS) include:
- photosynthesis
- transpiration
- infiltration

OUTPUTS include:
- evaporation
- seed dispersal
- animal migration
- logging
- drainage through soil and rocks
- nutrient leaching

🔺 **Figure 1** *Systems in a deciduous forest ecosystem in the UK*

Energy flows

The source of all energy is the Sun. It is captured by the green pigment in plant leaves (chlorophyll) that converts carbon dioxide and water into their organic compounds through the process of photosynthesis. These so-called 'building-blocks' of plants are tissue and food energy in the form of chemicals called *carbohydrates*.

The total amount of energy absorbed or fixed by green plants is called the *gross primary productivity (GPP)*, some of which is lost through respiration. The remaining *net primary productivity (NPP)* is used in the production of leaves – the higher the value of light, warmth, water and nutrients, the greater the productivity (Figure **2**).

Producer	Biomass productivity (gC/m²/yr)
Tropical rainforests	2000
Coral reefs	2000
Temperate forests	1250
Cultivated lands	650
Tundra	140
Open ocean	125
Deserts	3

Biomass is the total weight of all biotic (living) organisms per unit area – usually expressed as the mass of organically bound carbon present.

🔺 **Figure 2** *Selected global producers of biomass productivity*

Continued over ▶▶▶

Food chains and webs

Carbohydrates contain all the organic materials needed by animals for growth, movement and reproduction (e.g. starches and proteins). Plants, therefore, form the basis of all nutrition and energy for the whole ecosystem. This is because they provide food for other organisms – which, in turn, feed others in what we call a *food chain* (e.g. plant → insect → toad → snake → fox).

However, just as animals have different sources of food, any one species of plant or animal will be eaten by a variety of different consumers – so resulting in complex networks of interconnecting food chains called *food webs*.

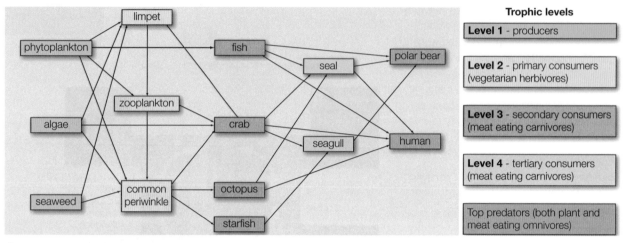

Trophic levels

| Level 1 - producers |
| Level 2 - primary consumers (vegetarian herbivores) |
| Level 3 - secondary consumers (meat eating carnivores) |
| Level 4 - tertiary consumers (meat eating carnivores) |
| Top predators (both plant and meat eating omnivores) |

🔺 **Figure 3** *A marine ecosystem food web*

Trophic levels and energy pyramids

Plants represent the first **trophic** (or energy) **level** in the food chain or web. These producer organisms (self-feeding *autotrophs*) produce their own food through photosynthesis. All other trophic levels are occupied by consumers which include all animals – including humans (Figure **3**).

Although nutrients are recycled in ecosystems, 90% of energy is lost at each trophic level by animal respiration, movement and excretion. So, fewer organisms can be supported at each level.

Energy pyramids show this well and have been used to demonstrate the extravagant food and energy wastage, in terms of processing, involved in diets found in HDEs compared with LDEs (Figure **4**).

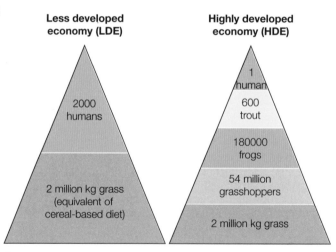

🔺 **Figure 4** *In an LDE the equivalent of 2 million kg of grass feeds 2000 people; in an HDE the same amount feeds just one person*

You need to know:
- the definitions of nutrients and the nutrient cycle
- what the Gersmehl Nutrient Cycle is.

Student Book
pages 294–5

What is a nutrient cycle?

Nutrients are plant foods consisting of minerals and chemicals derived from:

- precipitation
- rock weathering
- decomposition of plant matter.

The **nutrient cycle** constantly recycles the vital nutrients between the soil and the plants – essential to sustain life in an ecosystem. It is commonly represented in a standard format developed by P.F. Gersmehl in 1976 (Figure **1**). This takes the form of a simple system:

- *Stores* – biomass, litter and soil represented by proportional circles. The bigger the circle, the greater the store's importance.
- *Inputs* – two key external inputs, namely:
  - nutrients dissolved in precipitation
  - weathered parent material (e.g. bedrock) dissolved in soil water.
- *Outputs* – nutrients lost from the system by runoff and leaching.
- *Transfers* – littering (e.g. leaf drop), decomposition and plant uptake from the soil represented by proportional arrows. The thicker the arrow, the greater its significance.

The Gersmehl Nutrient Cycle

The Gersmehl Nutrient Cycle is usually used for comparing widespread variation between nutrient stores and transfers across major terrestrial biomes. For example, warm, wet climatic conditions in a tropical rainforest dictate the:

- larger biomass store (of dense, lush vegetation) compared to a grassland (prairie)
- greater nutrient transfer from rapid parent rock weathering compared to little in a coniferous forest
- minimal litter store owing to rapid decomposition.

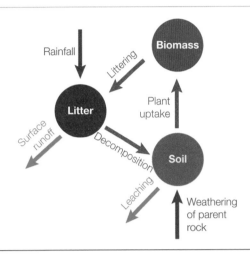

Arrows; inputs, outputs, transfers

Nutrient inputs into the ecosystem
- Nutrients dissolved in raindrops
- Nutrients from weathered rock

Nutrient outputs (losses) from the ecosystem
- Nutrients lost through surface runoff
- Nutrients lost through leaching

Nutrient transfers within the ecosystem
- Littering – fallout from the plants, mostly leaf fall transferring nutrients to the litter
- Decomposition – decay of organic material in the litter by fungi and bacteria transferring nutrients to the soil
- Plant uptake – the uptake of nutrients from the soil through the plant roots

Circles: nutrient stores in the ecosystems
- Biomass – contains all living plant and animal matter in the ecosystem
- Soil – contains minerals from the parent rock in addition to humus from decomposed plant and animal remains
- Litter – sits on top of the soil and contains both dead and decaying plant and animal material

Figure 1 *The Gersmehl Nutrient Cycle*

- Plant nutrients consist of minerals and chemicals derived from precipitation, the weathering of rock and the decomposition of organic matter.
- The nutrient cycle (constantly recycling vital nutrients between the soil and the plants) is vital in sustaining life in an ecosystem.
- The Gersmehl Nutrient Cycle uses proportional circles and arrows to show the relative importance of nutrient stores contained in, and flows between the biomass, litter and soil.
- The nutrient cycles of major terrestrial biomes demonstrate that the overriding controlling factor is climate.

Over to you

Revise tropical rainforest (6.7) and savanna grassland (6.8) biomes before sketching and explaining Gersmehl Nutrient Cycles for each.

Student Book
pages 296–303

You need to know:
- the factors affecting terrestrial ecosystems and the interaction between them
- how terrestrial ecosystems respond to change
- the effect of alternative futures on terrestrial ecosystems.

What are terrestrial ecosystems?

There is a huge range of terrestrial ecosystems – from small-scale ecosystems (e.g. field margins, hedges and ponds), to larger-scale ecosystems (biomes) (e.g. deserts, tundra and tropical rainforests).

Interaction of factors affecting terrestrial ecosystems

Five factors determine the characteristics of an ecosystem. They are linked with one another.

- *Climate:* determines, to a large extent, which plants colonise a particular region.
- *Topography (Relief):* valley floors accumulate deeper, wetter soils than slopes. Altitude affects climate, which, in turn, affects vegetation.
- *Vegetation:* the primary food supply and provides habitats.
- *Soils:* depth, texture and fertility determine the type and abundance of plants.
- *Time:* soil formation and subsequent colonisation by vegetation is very slow.

Read about the interrelationships on Exmoor and in Sierra Nevada on pages 297–8 in the student book.

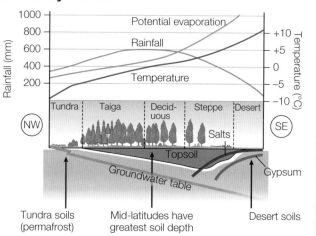

△ **Figure 1** *NW–SE profile through eastern Europe showing variations in climate, soils and vegetation*

How do ecosystems respond to change?

Component	Natural change	Anthropogenic (human) change
Climate	Natural climate change is generally slow enabling ecosystems to adjust. Short-term cycles (e.g. El Niño) can be more damaging to ecosystems by causing floods and drought.	Recent warming, causing changes to the seasons and gradual latitudinal shifts in vegetation belts, is affecting ecosystems. Extreme weather may be linked to climate change caused by human activity.
Vegetation	The vegetation succession results in changes to food supply and habitats. These changes affect wildlife but they happen very slowly, usually over hundreds of years.	The spread of invasive plant species and alien animal species can impact significantly on an ecosystem. Human management (drainage, rotational burning, deforestation) can be devastating to ecosystems.
Soil	Soil erosion or incidents of pollution could inflict immediate change on an ecosystem.	Indirectly, management of vegetation impacts on soils. They might become compacted or eroded, for example.
Topography	Topography usually develops extremely slowly, except where there are high rates of erosion or active mass movement processes.	Terracing of slopes or slope stabilisation affects drainage and soils which could impact on ecosystems.

△ **Figure 2** *This table outlines the main causes of ecosystem change*

Impacts of wildfires on ecosystems

Wildfires clear dead wood, releasing important nutrients (e.g. potash, phosphate) and can trigger seed germination (e.g. the Ponderosa pine tree). However, increasing frequency disrupts natural cycles and native species may suffer.

Wildfires can increase atmospheric carbon dioxide levels, strengthening the greenhouse effect and increasing the likelihood of fires in the future.

Impacts of conservation on ecosystems

Tree planting restores local ecosystems affected by deforestation. Trees affect water movement, soil development and the range of flora and fauna that can live and thrive in an environment. Other conservation measures include the restoration of peatlands and footpaths.

Read about invasive species on pages 299–300 in the student book.

How might ecosystems be affected by alternative futures?

Climate change

Changes to the climate, particularly rapid changes, have the potential to cause significant impacts, especially if species are unable to adapt at the same rate at which change occurs.

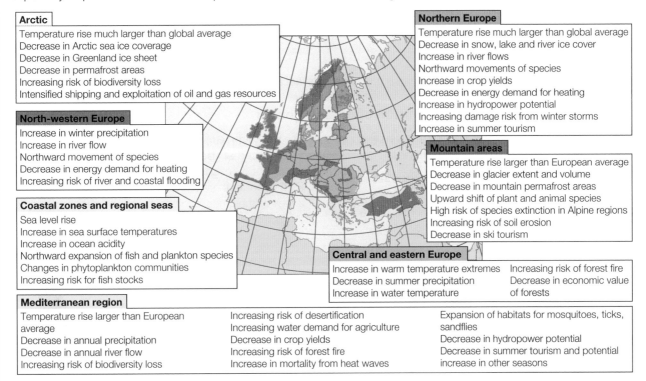

Arctic
Temperature rise much larger than global average
Decrease in Arctic sea ice coverage
Decrease in Greenland ice sheet
Decrease in permafrost areas
Increasing risk of biodiversity loss
Intensified shipping and exploitation of oil and gas resources

North-western Europe
Increase in winter precipitation
Increase in river flow
Northward movement of species
Decrease in energy demand for heating
Increasing risk of river and coastal flooding

Coastal zones and regional seas
Sea level rise
Increase in sea surface temperatures
Increase in ocean acidity
Northward expansion of fish and plankton species
Changes in phytoplankton communities
Increasing risk for fish stocks

Northern Europe
Temperature rise much larger than global average
Decrease in snow, lake and river ice cover
Increase in river flows
Northward movements of species
Increase in crop yields
Decrease in energy demand for heating
Increase in hydropower potential
Increasing damage risk from winter storms
Increase in summer tourism

Mountain areas
Temperature rise larger than European average
Decrease in glacier extent and volume
Decrease in mountain permafrost areas
Upward shift of plant and animal species
High risk of species extinction in Alpine regions
Increasing risk of soil erosion
Decrease in ski tourism

Central and eastern Europe
Increase in warm temperature extremes
Decrease in summer precipitation
Increase in water temperature
Increasing risk of forest fire
Decrease in economic value of forests

Mediterranean region
Temperature rise larger than European average
Decrease in annual precipitation
Decrease in annual river flow
Increasing risk of biodiversity loss
Increasing risk of desertification
Increasing water demand for agriculture
Decrease in crop yields
Increasing risk of forest fire
Increase in mortality from heat waves
Expansion of habitats for mosquitoes, ticks, sandflies
Decrease in hydropower potential
Decrease in summer tourism and potential increase in other seasons

🔺 *Figure 3* *Threats, challenges and opportunities of climate change in Europe*

Human exploitation of the environment

Figure **4** shows how coastal ecosystems have been affected by human exploitation.

Human exploitation by deforestation has cleared half of all mature forests.

Tropical rainforests are home to about 90% of the world's species, and it is these that are most threatened.

Goods & services
Food
Recreation
Tourism
Biodiversity
Trapping sediment
Coastal erosion control
Mining
Shipping
Aesthetic landscapes
Culturally important places

Coral reefs are being harmed irretrievably by overfishing, pollution and global warming.

The destruction of mangrove forests and salt marshes for development has massive impacts on flora and fauna.

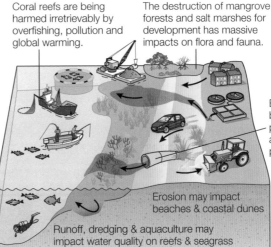

Ecosystems have been harmed by pollution, oil spills and waste, particularly plastics.

Erosion may impact beaches & coastal dunes

Runoff, dredging & aquaculture may impact water quality on reefs & seagrass

See page 302 of the student book for examples of the effects of climate change

🔺 *Figure 4* *This diagram shows the impact of human exploitation on terrestrial and marine coastal ecosystems*

Sixty second summary

- There are many terrestrial ecosystems (e.g. forests, grassland, reservoirs).
- Factors affecting terrestrial ecosystems include climate, topography, soils, vegetation and time.
- Terrestrial ecosystems respond to natural change and human actions.
- Terrestrial ecosystems may be affected by alternative futures (e.g. climate change and human exploitation).

Over to you

Figure **3** contains a wealth of valuable information. Highlight up to half of each text box to allow you focus on the key elements. Start with a pencil and only commit to a highlighter pen when you're certain.

Student Book
pages 304–5

You need to know:

- the definition of global biomes
- the distribution of global biomes
- the role of climate and relief in determining the features and distribution of biomes.

What is a biome?

A **biome** is a large-scale ecological area with plants and animals that are well adapted to their environmental conditions.

The role of climate in the distribution of global biomes

The pattern of global biomes more or less reflects the pattern of global climate zones (Figures **1** and **2**).

- The high rainfall and constant high temperatures in the equatorial regions create ideal conditions for tropical rainforests, hence their broad swathe across South America, Africa and into south-east Asia.

- The very high temperatures and low rainfall experienced at about 30 °N and 30 °S equates with the major world deserts.
- Further north, there are close links between climate zones and biomes, such as the Mediterranean region, the European temperate and coniferous forests and the sub-Arctic tundra.

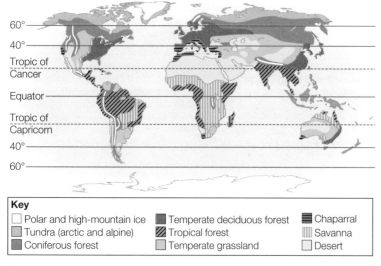

Key

☐ Polar and high-mountain ice	▨ Temperate deciduous forest	▤ Chaparral
▨ Tundra (arctic and alpine)	▨ Tropical forest	▥ Savanna
▨ Coniferous forest	▨ Temperate grassland	☐ Desert

🔺 *Figure 1* *The global distribution of terrestrial biomes*

🔺 *Figure 2* *The effect of global biomes on temperature and precipitation*

The role of relief (altitude) in the distribution of global biomes

Relief (altitude) is an important factor affecting biomes (Figure **3**). As altitude increases, so temperature falls and rainfall tends to increase. This results in a zonation of vegetation (biomes) with altitude mirroring the latitudinal changes.

Figure 3 🔵

Increasing altitude and latitude have these effects on global biomes

Sixty second summary

- Biomes are large-scale ecological areas, with well-adapted plants and animals.
- Climate is the main factor affecting the distribution and characteristic features of biomes.
- Relief has a significant effect on temperature which leads to a clear biome succession with increasing altitude.

Over to you

Make a list of the different biomes from the Arctic to the Equator and link **each one** to temperature and rainfall.

You need to know:

- about the characteristics and controlling factors of tropical rainforests
- the adaptations by flora and fauna
- the issues associated with human activity.

Student Book
pages 306–11

Characteristics, distribution and adaptations

The tropical rainforest biome is mostly located within the equatorial climate belt, an area 5° either side of the Equator (Figure **1**). Tropical rainforests now cover less than 5% of the Earth's surface, yet they still support 50% of the planet's living organisms – forming close **symbiotic** relationships with each other and evolving effective **adaptations** (Figure **2**).

- *Equatorial climate* has no distinct seasons, is consistently hot (average daily 28 °C) and humid all year, with high, but variable rainfall (around 2000 mm).
- *Biodiversity* is remarkable, supporting more plants and animals than any other biome.
- *Vegetation* is perfectly adapted to the environment, with distinctive vertical stratification of deciduous trees and climbing plants, which form a dense canopy all competing for light.
- *Soils* are iron-rich *laterites* – fragile and surprisingly infertile because *nutrient cycling* is so rapid. Once vegetation is cleared the soils are prone to rapid *leaching*.

 Big idea

Luxuriant, dynamic and vulnerable, tropical rainforests inspire stronger passions, more superlatives and expressed concerns than any other biome on Earth.

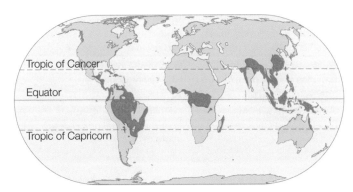

▲ **Figure 1** *Distribution of tropical rainforests*

Emergent layer (35–50 m) Enormous, exposed, hardy trees with straight branchless trunks receive the most light.

Adaptations include fluffy kapok tree seeds carried by the wind.

Understorey (10–20 m) Shaded, slender, less substantial trees with thin, smooth bark waiting their chance to take advantage of the next light space to become available. Interlocking spindly branches and climbing woody creepers (*lianas*) form green corridors along which lightweight animals can travel.

Adaptations include bats living in the low light.

Canopy (20–35 m) The most heavily populated and productive layer as each crown has an enormous photosynthetic surface of dark, leathery leaves.

Adaptations include *drip tips* which help them shed water quickly and efficiently, and so prevent rotting.

Shrub layer (2–10 m) Lack of light limits vegetation to ferns, woody plants and younger trees.

Adaptations include omnivores developing survival strategies of speed, stealth or camouflage.

Forest floor (0–2 m) Bacteria and fungi rapidly rot the fallen leaves, dead plants and animals.

Adaptations include thick *buttress* roots that spread the weight of the towering trees above.

▲ **Figure 2** *Stratification of vegetation and flora and fauna adaptations in a tropical rainforest*

Continued over ▶▶▶

Human activity and deforestation in tropical rainforests

The effects of human activity – most notably tropical rainforest deforestation – are widespread and affect us all. Rapid tropical rainforest deforestation has economic, environmental and social impacts (Figure **3**).

Cause	Details	Impacts	Examples
Farming: Agricultural land needed for rapidly growing LDE populations.	Space also needed for *plantations* of cash crops (e.g. soya) and *cattle ranches*.	*Cattle ranching* exposes pastures to *soil erosion* and *leaching*.	Soya grown in Amazonia is used to feed chickens in the UK.
Logging of valuable equatorial hardwoods (e.g. mahogany) provides a reliable and essential source of income.	HDE demand for furniture and building materials has encouraged felling.	Destructive *clear felling* exposes the ground and decimates animal habitats and species (e.g. orangutans).	Japan alone accounts for a staggering 11 million cubic metres of equatorial hardwood used a year.
Mining of abundant, valuable minerals (e.g. iron ore, copper and gold).	Cheapest method of mineral extraction is *opencast* mining.	Large-scale deforestation as trees and soil are stripped. Silted water courses lead to *flooding*.	Carajás in northern Brazil is the world's largest source of iron ore.
Road construction supports the development of rainforest for other uses.	New private and government-funded roads allow people in and raw materials out.	Broad swathes cut across the rainforest – *habitat destruction*.	The Trans-Amazonian Highway extends 6000 km into Brazil's interior.
Hydro-electric power (HEP) provides cheap, plentiful, renewable energy.	High rainfall gives potential for HEP generation.	Reservoirs flood large areas of cleared forest.	HEP station on the Tocantins River powers Carajás mine.
Settlement growth reflects the rapid population increase.	Population pressure comes from both natural growth and migration.	Brazil has over 25 million landless people – relocation cannot be ignored if other areas cannot cope.	Migration from the poorest parts of Brazil (e.g. drought-stricken north-east) encouraged.

▲ *Figure 3* Causes and impacts of tropical rainforest deforestation

In the Amazon alone, 17% of the tropical rainforest has been lost in the last 50 years, almost half the size of continental Europe!

- Deforestation poses a huge threat to biodiversity and the long-term sustainability of the tropical rainforest biome.
- Animal species have been decimated. In Malaysia, more than half of river fish species have disappeared due to logging.

- Removal of the protecting canopy exposes soils to direct rainfall leading to soil erosion and leaching. More sediment reaches water courses leading to localised flooding and more soil erosion. Reduced rates of evapotranspiration may change regional rainfall patterns.
- A tropical tree acts as a carbon sink. Deforestation accounts for around 20% of global atmospheric carbon dioxide released each year, mostly through burning.

Sixty second summary

- The tropical rainforest biome lies within the equatorial climate belt, 5° either side of the Equator.
- Tropical rainforests now cover less than 5% of the Earth's surface, yet still support 50% of the planet's living organisms.
- This is the Earth's most productive biome, with the richest biodiversity.
- The equatorial climate creates ideal growing conditions.
- Despite fragile, iron-rich laterite soils, flora and fauna adapt and thrive within the stratified vegetation layers, forming close symbiotic relationships.
- Rapid tropical rainforest deforestation (for agriculture, logging, mining, road construction, HEP and settlement growth), has economic, environmental and social impacts.

Over to you

Check your understanding of the word 'biodiversity', and consider what plants, animals and adaptations you would use to illustrate it in the tropical rainforest.

Student Book
pages 312–17

You need to know:

- characteristics and distribution of the savanna biome
- about climate, vegetation and soils in the savanna biome
- how flora and fauna have adapted and impacts of human activities.

Distribution of savanna grassland

Savanna grassland covers around a third of the Earth's land surface, within the tropics roughly between 15 °N and 30 °S (Figure **1**). The greatest concentrations are in Africa, South America, parts of south-east Asia and Australia and lie mainly between tropical rainforest and desert.

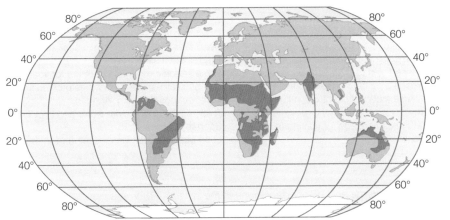

🔺 **Figure 1** *The distribution of savanna grasslands*

Characteristics of savanna grassland biome

Savanna is characterised by vast grasslands with scattered shrubs and isolated trees, with insufficient rainfall to support rainforest. These landscapes are often associated with animals such as giraffes, elephants and lions.

Climate	• Clearly defined wet seasons (May to September in the northern hemisphere and November to March in the southern hemisphere). • The reliability and amount of rainfall decreases with distance from the Equator. • There are more trees in the wetter areas close to the Equator and more grassland in the drier regions away from the Equator. • Fires are common in the drier areas, often ignited by lightning strikes. They are an important input to the system, burning off dead vegetation and often stimulating new growth.
Vegetation	• The predominant vegetation is grassland. • Grass is ideally suited to the conditions (see Figure **4**). • Vegetation varies considerably (e.g. in Africa, the acacia tree is commonly associated with this biome, in northern Australia, it is the eucalyptus).
Soils	• Porous soils due to grass root penetration, so the soils are well drained during the wet season. • Porous soils aid leaching (a transfer), resulting in infertile topsoil – any humus created remains close to the surface (Figure **3**). • Leaching causes characteristically red soils (*laterites*) due to the presence of iron oxide (Figure **3**). • The soil water budget fluctuates – a surplus in the wet season and a deficit in the dry season.

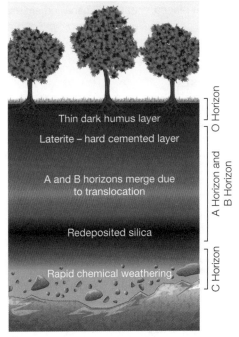

🔺 **Figure 3** *Typical savanna grassland soil profile*

🔺 **Figure 2** *Climatic, vegetation and soil characteristics of the savanna biome*

Continued over ▶▶▶

Adaptations of flora and fauna in savanna grasslands

Plants and animals cope with high temperatures throughout the year, as well as periods of intense rainfall followed by long periods of drought and associated wildfires. Plants have to cope with infertile soils and a variable soil water budget, reflecting the seasonal nature of the rainfall.

Some animals that live on the savanna live off the grass, and others eat leaves from trees. They eat different plants at different times of day and in different areas. This interspecies survival is the key to retaining biodiversity and sustainability.

Animal adaptations	Plant adaptations
• Long legs enable long migrations to follow the rains. • Some animals live in burrows to avoid the heat of the day. • Some lose heat through large areas of exposed skin (e.g. ears of elephants). • Birds of prey nest in isolated trees, giving excellent field of vision to spot prey. • Natural camouflage for hunting (e.g. hyenas).	• Grasses are dormant in the dry season and then regrow quickly in the wet season. • Wiry grasses turn their blades away from the Sun to reduce water loss. • Small, waxy leaves or thorns to reduce water loss. • Deep taproots enable them to seek water deep underground. • Fire-resistant stems. • Some seeds can remain dormant for years to survive prolonged drought. • Low umbrella-shaped canopies provide shade for the shallow roots, reducing soil water evaporation (e.g. Acacia). • Thick, fleshy trunks store water and thick barks insulate and protect from fire (e.g. Baobab). • Long taproots seek water deep underground.

Figure 4 *Animal and plant adaptations in the savanna biome*

The impacts of human activity in savanna grasslands
Overgrazing and ranching

Agriculture, in particular overgrazing, is threatening savanna grasslands. Traditionally, nomadic herders have driven their domestic animals (cattle, sheep and goats) across the grasslands with little damage. However, overgrazing can be a real problem with more permanent agriculture, especially if exacerbated by drought and, in the future, by climate change.

In some places, grassland management uses fire to clear areas of dead vegetation and stimulate fresh growth of grass for grazing. Such fires can kill young shrubs and trees, reducing biodiversity and the range of habitats.

Overgrazing can be considered to be a positive feedback loop – it reduces the amount of vegetation, increasing the pressure on the remaining grassland, which, in turn, leads to more overgrazing and soil erosion.

Tourism

Tourism, particularly safaris, can lead to high concentrations of people in small areas and development pressures, such as new roads, runways and hotels. Animals can become scavengers, eating human rubbish and being poisoned by toxic substances. The intensive use of vehicles can damage the grass, leading to soil erosion.

 Sixty second summary

- The savanna biome is generally sandwiched between tropical rainforest and desert biomes.
- Distinctive wet and dry seasons accompany high temperatures. Wildfires are common.
- Grass lies dormant during the dry season and grows quickly during the wet season.
- Soils tend to be porous and red in colour due to leaching.
- Plants and animals are well adapted to the conditions.
- Overgrazing and tourism can damage fragile ecosystems, leading to soil erosion and desertification.

 Over to you

On flashcards headed, 'characteristics', 'distribution', 'climate', 'adaptations' and 'impacts', summarise the key features of the savanna biome.

Student Book
pages 318–25

You need to know:

- about vegetation succession and climatic climax in the British Isles
- the impact of human activity on vegetation succession.

Vegetation succession (seral progression)

Plant communities vary from area to area, evolving and becoming more complex over time – this is *vegetation succession*. If allowed to continue undisturbed, each distinctive *seral* stage in the progression will climax in a state of perfect adaptation to the environment at the time. Climate is often the main controlling factor, so this state of equilibrium is sometimes called the **climatic climax vegetation** (this is deciduous woodland in the UK).

There are two types of vegetation succession.

- Primary successions (*priseres*) occur on any surfaces that have had no previous vegetation, e.g. vegetation succession on sand dunes – a psammosere, or on bare rock – a **lithosere** (Figure **1**).
- Secondary successions (*subseres*) occur on surfaces that have already been colonised but subsequently modified or destroyed (e.g. by deforestation).

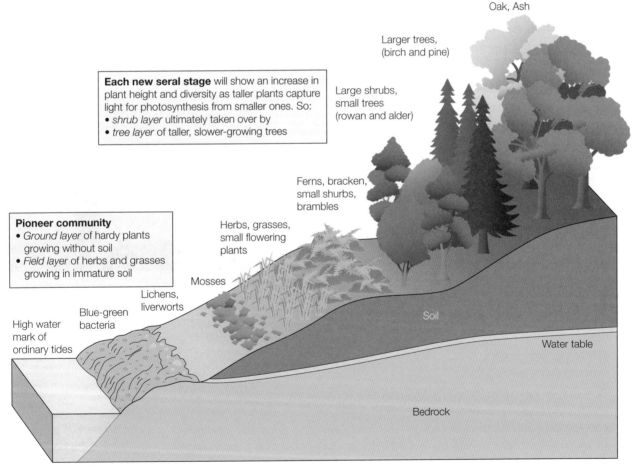

Climatic climax
With no environmental changes the tallest trees will prevail in a state of equilibrium until circumstances change

Oak, Ash

Larger trees, (birch and pine)

Each new seral stage will show an increase in plant height and diversity as taller plants capture light for photosynthesis from smaller ones. So:
- *shrub layer* ultimately taken over by
- *tree layer* of taller, slower-growing trees

Large shrubs, small trees (rowan and alder)

Ferns, bracken, small shurbs, brambles

Herbs, grasses, small flowering plants

Pioneer community
- *Ground layer* of hardy plants growing without soil
- *Field layer* of herbs and grasses growing in immature soil

Mosses

Lichens, liverworts

Blue-green bacteria

High water mark of ordinary tides

Soil

Water table

Bedrock

🔺 **Figure 1** *Natural vegetation succession along a raised beach – a lithosere*

📖 Examples of vegetation successions are on pages 319–21 of the student book.

Continued over ▶▶▶

Temperate deciduous woodland biome

If there was no agriculture, urbanisation or infrastructure, temperate deciduous woodland would be the climatic climax vegetation in Britain. Isolated pockets remain, but most represent secondary successions.

Deep, fertile, well-drained *brown earth* soils support a productive, high-energy biome thriving in favourable growing conditions – plenty of rainfall, warm summers and cool winters. The autumn leaf shed helps trees conserve water loss through transpiration, whilst protecting them from frost and snow (Figure **2**). Seasonal leaf cover also affects the levels of light and humidity in the woodland – accounting for distinctive stratification:

- *Ground layer* – shaded mosses and lichens among thick leaf litter, ideal for bacteria and fungi to thrive.
- *Field layer* – spring and early summer brambles, bracken and bluebells, flourish before the leaf canopy above has fully developed.
- *Shrub layer* – low bushes and smaller trees, such as rowan and hawthorn, all compete for light.
- *Canopy layer* – oak and other tall deciduous species, such as beech and ash, with extensive crowns of broad leaves which absorb maximum sunlight during the summer.

Temperate woodlands are extremely biodiverse. They support a large range of fauna that has adapted by migrating (e.g. birds), hibernating (e.g. hedgehogs), storing food for winter (e.g. squirrels) or changing colour (e.g. stoats).

⚠ **Figure 2** *Temperate deciduous woodland, Standish Wood, Gloucestershire*

The effects of human activity on succession

Human activity frequently impacts on plant succession. The resulting human-induced vegetation climax is referred to as a *plagioclimax*. For example:

- 3000 years ago the North Yorkshire Moors were covered in climatic climax deciduous woodland on fertile *brown earth* soils.
- This was cleared for farming, exposing the ground to heavy upland rainfall.
- Erosion and leaching left thin, acidic, *podsols*, which only more hardy plants could tolerate.
- This moorland vegetation is now managed using controlled burning.
- This maintains heather for the highly lucrative (red grouse) shooting industry.

⚠ **Figure 3** *Controlled burning keeps as much of the heather moorland as possible in the most productive 'building phase'*

 Sixty second summary

- The development and associated changes in a plant community through time are known as vegetation succession.
- All vegetation successions pass through a series of seral stages, starting with the pioneer community (ground layer), followed by progressively taller field, shrub and tree layers.
- Providing that environmental factors do not change, or human activities intervene, natural vegetation succession should eventually reach a state of equilibrium – the climatic climax vegetation.
- The highly productive temperate deciduous woodland biome is the natural climatic climax vegetation of the British Isles.
- An artificial human-induced vegetation climax is known as a plagioclimax community (e.g. heather moorland).

Over to you

Make sure you are familiar with and understand the key terms of this topic (identified in **bold** and also in *italics*).

Student Book
pages 326–33

You need to know:

- the characteristics and importance of coral reefs
- factors in the development of coral reefs, including nutrient cycling
- what the threats are to coral reef ecosystems
- the impacts of human activity.

Distribution of coral reefs

Warm coral reefs (there are coral that survive in cold climates) need a delicate balance of marine environmental conditions to survive and thrive. They are mainly found in the tropics (Figure **1**).

Characteristics and importance of coral reefs

Coral reefs are formed by the hard exoskeletons of millions of tiny coral animals. They account for only about 18% of the marine environment yet they are home to nearly 25% of all marine species.

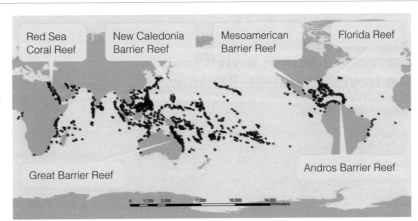

Figure 1 *Most of the world's coral reefs are found in the tropics*

- They form one of the richest ecosystems on Earth.
- More than 450 million people live within 60 km of coral reefs, deriving food and income from them.
- In the Caribbean alone, coral reefs generate some US$10 billion annually.
- Coral reefs form a natural barrier to protect the mainland from storms.
- Algae and sponges on coral reefs have valuable medicinal qualities.
- Coral reefs store carbon (calcium carbonate) – important in the carbon cycle.

Temperature	Need an average temperature above 18 °C; ideally 23–25 °C.
Salinity	Need saline water, so there are gaps in reefs at freshwater river mouths.
Acidity	Need alkaline water; increased acidity (e.g. increased CO_2) can kill them.
Clear water	Sediment clogs corals' feeding structures. It also reduces the amount of light; another reason why coral reefs are not found at river mouths.
Air	Exposure to air kills coral; so upward growth is limited to the level of the lowest tides.
Light	Corals feed on algae, which need light to photosynthesise and grow; so coral reefs are found in relatively shallow water, with enough light for algae to thrive.

◀ **Figure 2**
Environmental conditions needed for coral reef development

Nutrient cycling and the role of algae in coral reefs

Corals live in nutrient-poor waters, so efficient nutrient recycling is essential to maintain such a diverse ecosystem. At the heart of the recycling is a symbiotic relationship between coral and algae which captures nitrogen effectively. Corals are also able to digest *zooplankton*, bacteria and edible detritus that enter the system by upwelling from the ocean floor.

- *Zooxanthellae* (plant-like algae) live within the coral *polyp* (the tiny animals that make up the coral). They convert sunlight into energy to provide nutrients.
- The zooxanthellae benefit by having access to the coral's waste-nutrients (nitrogen and phosphorus), which fertilise the algae.

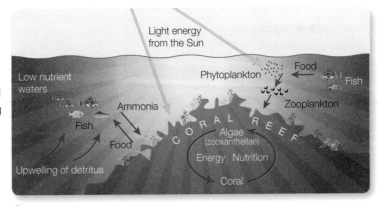

Figure 3 *Despite the nutrient-poor waters, a complex ecosystem thrives due to efficient nutrient recycling*

Continued over ▶▶▶

What are the threats to coral reef ecosystems?

- Urban development or deforestation can wash silt into the sea, clogging the corals' feeding mechanism; or change the salinity of the water.
- Nutrient-rich agricultural and sewage discharges can lead to the growth of algal blooms which smother corals and block sunlight.
- Physical damage associated with fishing or tourism.
- Climate change can lead to acidification and coral bleaching – a huge threat, which could destroy vast swathes of coral reef in the future.

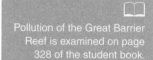

Pollution of the Great Barrier Reef is examined on page 328 of the student book.

Impacts of human activity: Andros Barrier Reef, Bahamas

Climate change

Carbon dioxide reacts with seawater and increases acidity, which can dissolve the calcium carbonate corals that molluscs, such as crabs and lobsters, need to build their shells and stony skeletons.

Higher water temperatures can trigger a stress reaction in corals, causing them to expel the zooxanthellae. This causes the coral to become 'bleached', literally turning white. Eventually the coral dies.

Overexploitation

Commercial fishing and tourism can cause immense harm to the coral reef ecosystem.

- The fishing of grouper on the Andros Barrier Reef has led to an increase in damselfish, upon which groupers feed. Damselfish, in turn, create habitats in coral for algae, which can smother a reef.

- Overfishing of plant-eating fish can also lead to increased algal growth.
- Corals can be killed by physical contact with anchors, fishing nets, boat hulls and even people's feet.
- The harvesting of sponges can imbalance the ecosystem.

Pollution

- Silt restricts the penetration of sunlight used by zooxanthellae to photosynthesise.
- Clearing of vegetation for coastal developments has increased soil erosion and silt deposition.
- Marine-based oil and chemical pollution from trawlers and other ships can be harmful.

Hurricanes

Coral can be scoured by powerful waves and damaged by the snapping of branching corals and removal of sponges. Storms can also disturb seabed sediments, clouding the water and potentially clogging up the corals' feeding systems.

What are the future prospects for coral reefs?

In 2011 the World Resources Institute reached the following conclusions.

- More than 60% of the world's coral reefs are under local threat from overfishing/destructive fishing, coastal development, watershed- or marine-based pollution.
- This increases to 75% if thermal stress is considered (ocean warming linked to climate change).
- If left unchecked, this will increase to 90% by 2030 and to nearly all reefs by 2050.

 Sixty second summary

- Coral reefs are extremely biodiverse, supporting thousands of species of organisms and benefiting people (food supply, tourism, medicines, etc).
- Corals are found in warm, clear, pollution-free tropical waters.
- Nutrient cycling between corals and algae (zooxanthellae) is essential in the nutrient-poor waters.
- Threats include land degradation, intensive agriculture, tourism, oil exploitation and climate change.
- Andros Barrier Reef (Bahamas) is one of the world's most biodiverse and healthy reefs.

 Over to you

Explain why coral reef ecosystems are so important.

You need to know:

- about the characteristics of urban wasteland ecosystems
- the factors affecting their development
- the impacts of change.

Student Book
pages 334–9

Characteristics of urban wasteland ecosystems

Urban wasteland means the abandoned *brownfield* sites essentially left for nature to 'take its course'. They include former factories or routeways (e.g. railway lines), dumping sites for industrial waste, building demolition sites and quarries. The variety of surfaces (e.g. bare soil and piles of rubble) and varied topography (e.g. hummocks, holes and wide open spaces) provides lots of different microhabitats.

Despite the particular difficulties that these sites pose (e.g. contaminated ground and limited soil depth) lithosere-type plant succession is surprisingly adaptive and rapid (Figure **1**).

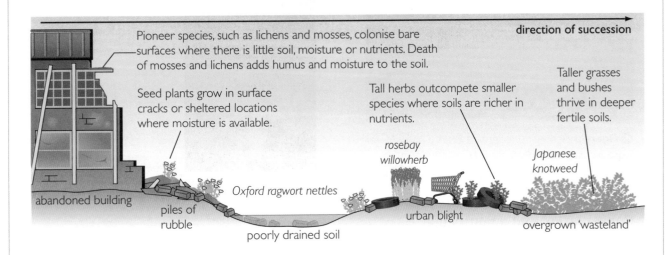

Pioneer species, such as lichens and mosses, colonise bare surfaces where there is little soil, moisture or nutrients. Death of mosses and lichens adds humus and moisture to the soil.

Seed plants grow in surface cracks or sheltered locations where moisture is available.

Tall herbs outcompete smaller species where soils are richer in nutrients.

Taller grasses and bushes thrive in deeper fertile soils.

direction of succession

rosebay willowherb

Japanese knotweed

abandoned building

piles of rubble

Oxford ragwort nettles

poorly drained soil

urban blight

overgrown 'wasteland'

🔺 **Figure 1** *Plant succession on abandoned industrial sites or wasteland*

Factors affecting the development of an urban wasteland ecosystem

There are several important factors that affect the development of an urban wasteland ecosystem:

- *Climate*: subtle variations in microclimate, such as sheltered sun traps and exposed, windy areas will be exploited by different flora and fauna, e.g. mosses in damp areas, spiders in drier areas, and flowering plants such as buddleia colonising sunny walls and attracting butterflies (Figure **2**).
- *Soils*: soils will be largely absent, thin and/or infertile – suitable for mosses and lichens. Where soils develop in cracks, deeper-rooted plants will flourish and even trees, such as laburnum.
- *Soil water budget*: urban wasteland sites are often dry and dusty, resulting in low moisture retention and negative soil water budgets, with potential evaporation exceeding precipitation. Plants adapt to these conditions in various ways (e.g. the shallow roots of rosebay willowherb enabling them to obtain moisture quickly following rainfall events).

🔺 **Figure 2** *Purple-flowering buddleia on waste ground*

Continued over ⟫⟫⟫

Issues of change in an urban wasteland ecosystem

Urban environments are subject to constant economic, social and political change, which creates urban wastelands. However, as circumstances change, so wastelands themselves become threatened. They may be:

See pages 337–9 in the student book for case studies of urban wasteland.

- wiped out to make way for new developments
- transformed by the introduction of invasive, alien (non-native) species
- conserved and managed to benefit local communities or to protect endangered species.

Urban redevelopment

Development may completely destroy the wasteland ecosystem. For example, ponds are drained or infilled, land is flattened and bulldozed, and vegetation removed with waste materials (Figure **3**).

🔺 **Figure 3** Removing a derelict factory and clearing wasteland at High Wycombe, Buckinghamshire

Invasive plant species

Invasive alien plant species, such as Japanese knotweed, can have a devastating impact (Figure **4**). It can only be effectively treated by spraying with strong herbicides and then burning the weed and roots.

◀ **Figure 4** Alien species like Japanese knotweed interrupt natural succession and restrict biodiversity by out-competing native species

Conservation and amenity use

In the UK, government policy focuses on targeting brownfield sites for new developments within urban areas, not considering the often thriving ecosystem that has already developed. This conserves greenfield sites outside the urban area.

Several pressure groups and charities such as Buglife, the Wildlife Trust and the RSPB have been active in the identification of brownfield sites that require conservation rather than development.

As a result, several sites have been protected (e.g. Canvey Wick, Essex was the first brownfield site to be designated an SSSI).

Increasingly, areas of urban wasteland are being claimed by local communities as valuable ecological assets – for leisure, recreation, education and even as urban farms (e.g Prinzessinnengärten, Berlin, Germany).

 Sixty second summary

- The varied topography and variety of surfaces and materials provides many different microhabitats.
- Despite contamination, limited soil depth and lack of both moisture and nutrients, lithosere-type plant succession is rapid.
- Derelict buildings, and piles of rubble and industrial waste, form a range of habitats for insects, birds and animals.
- Microclimates, soils and soil water budgets affect the development of urban wasteland ecosystems.
- Urban wasteland ecosystems are subject to change, including clearing for development and invasion by alien plant species.

 Over to you

Make sure that you can exemplify this topic with appropriate, specific and located references (e.g. Canvey Wick, Essex as the first brownfield site to be designated an SSSI).

160 **Chapter 6** – Ecosystems under stress

You need to know:

- the ecological characteristics of Exmoor National Park
- causes of ecological change
- responses to ecological change.

Case Study

Exmoor National Park and its ecological characteristics

Designated a national park in 1954, Exmoor is an area of upland moorland covering about 700 km² of west Somerset and north Devon in south-west England (Figure 1).

- There is a wide range of natural habitats (and associated flora and fauna) on Exmoor including heaths, coastal marshes, ancient woodlands and upland peat wetlands (*mires*).
- About 25% of Exmoor is uncultivated heath and moorland mainly used for sheep grazing or recreation.
- The *mires* of upland Exmoor are of great ecological importance and are rich in biodiversity.

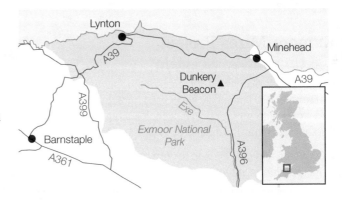

▲ *Figure 1* Location map of Exmoor National Park

Causes of ecological change on Exmoor

There are three main drivers of ecological change on Exmoor – agriculture, tourism and climate change.

Agriculture

Historically, agriculture has posed the greatest ecological threat on Exmoor. Farmers have dug drainage ditches in the peat moorlands to dry them out and make it more productive. This has made them more vulnerable to soil erosion by wind and rain.

Drying also diminishes water quality and sequestering of carbon from the atmosphere, and reduces biodiversity.

Currently, farming is dominated by sheep and beef cattle with the higher moors generally only used for rough grazing. Much of the low-lying land is cultivated.

Diversification by farmers into tourism has increased.

Tourism

It is estimated that some two million people visit Exmoor each year. Tourism brings in over £105 million to the economy and the equivalent of over 2000 full time jobs.

However, tourism can lead to ecological damage, particularly at honeypot sites, roadsides and car parks. Footpath erosion destroys vegetation, which leads to soil erosion and increased rates of surface runoff. Exmoor's roads are narrow and windy, resulting in congestion and occasional conflicts.

Climate change

Climate change may result in warmer summers and a longer growing season. This may increase productivity of commercial woodlands in Exmoor National Park.

However, the effects of climate change will have impacts.

- Some species are likely to find the climate tolerable (e.g. such as oak, beech, sweet chestnut), but some may suffer stress (e.g. common alder, small-leaved lime, black walnut).
- Upland woodland may become more like parts of lowland England with a greater mixture of broadleaf tree species.
- Many conifer species should grow well, although some (e.g. larch) may find it difficult to adapt.
- Woodland flora would be affected (e.g. lichen and ground flora), and may reduce in extent.
- There could be a greater risk of drought and woodland fires.
- Non-native pests and diseases may be able to survive milder winters and extend their range, putting native trees and woodlands under further stress.

Continued over ▶▶▶

Responding to ecological change

In 2011 Exmoor National Park Authority (ENPA published a collaborative study, *Exmoor Moorland Units*, that considered the issues, opportunities and management strategies within the National Park. The study involved local landowners as well as organisations such as the Forestry Commission, Natural England and the RSPB. Community involvement and support was considered essential.

Exmoor Mires Project

The Exmoor Mires Project (EMP) is an integrated management plan that addresses deteriorating peatlands on Exmoor. It is financially supported by South West Water and works with partner organisations such as ENPA, Natural England and the Universities of Exeter and Bristol.

The overall aim of EMP is to restore the hydrological function of the peatlands in the upper catchment of the River Exe by keeping the peat wet in order to withstand the effects of climate change.

After its launch in 2010, EMP conducted scientific research involving mapping, remote sensing and data collection. This enabled restoration plans to be drawn up for each site containing details of ecology, historic environment, landscape, access, land management and areas of ditch-blocking.

- Subsequent action has largely involved blocking the drainage ditches constructed in the past by farmers. This has reduced surface runoff and has resulted in the rewetting of the peatland.
- Restoration work is carried out between August and April to avoid the ground nesting bird season. Untreated timber is used to block the drains; in time this will biodegrade naturally, by which time the ditches will have silted up and peat will be re-establishing.

Between 2010 and 2015, 133 km of ditches were successfully blocked and 1139 ha of peatlands restored. Researchers at the University of Exeter found that there was a marked increase in species richness and diversity of mire flora and fauna
(see 1.18).

The role of Exmoor National Park Authority (ENPA)

ENPA has to strike a balance between conservation and supporting the community's economy, particularly its farmers.

It must also promote access for leisure and recreation. With tourism representing an increasing proportion of the region's income, ENPA has to work strategically with local and national political bodies (including local authorities and county councils) as well as national organisations, such as the RSPB and English Nature. ENPA does not own the land, it manages it. It, therefore, needs to work closely with landowners, as well as planning authorities and pressure groups.

This need for political cooperation and collaboration in the management of Exmoor's unique ecological setting is critical in promoting and supporting both conservation and continued economic development.

 Figure 2 *Blocking ditches to restore peatlands on Exmoor*

📖 Read about management of Dunkery Beacon on page 342 of the student book.

 Sixty second summary

- Exmoor National Park has a wide range of habitats (e.g. moorland, peat wetlands) and a rich cultural and ecological heritage.
- Agriculture, tourism and climate change are the main drivers of change.
- There have been a number of management responses, primarily involving the Exmoor National Park Authority and South West Water (Exmoor Mires Project).

Over to you

Outline responses to regional ecological change on Exmoor.

You need to know:

- why Wicken Fen is a biodiversity hotspot
- challenges, opportunities and management of Wicken Fen.

Where is Wicken Fen?

Wicken Fen is a 758 ha wetland nature reserve, 15 km to the north-east of Cambridge (Figure **1**). This unique landscape of natural lakes, meadows and fen wetlands has been managed by the National Trust since 1899.

Despite its conservation status, Wicken Fen is facing development threats from housing, transport and commercial agriculture, much of it driven by one of the fastest-growing cities in the UK, Cambridge.

A biodiversity hotspot

Wicken Fen comprises multiple habitats (scrub, sedge fields, woods, reeds, ponds and ditches) each with its own distinct ecosystem – a 'biodiversity hotspot'. *Lodes* (straight, raised Roman waterways) support different habitats that interlink with each other.

- The slow-flowing, clear waters provide perfect conditions for submerged aquatic plants to photosynthesise.
- Herbivores (e.g. water snails) feed on the aquatic vegetation.
- Small carnivorous fish (e.g. roach, minnows) are prey for larger carnivores (e.g. pike).
- Some species (e.g. dragonflies) move between aquatic and terrestrial ecosystems, forming an important component in two food webs.

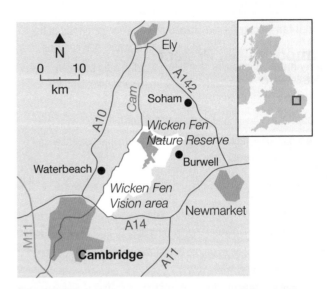

Figure 1 *Location of Wicken Fen Nature Reserve and Wicken Fen Vision area, 'one of the most important wetlands in Europe'*

Figure 2 *Wicken Fen – a 'biodiversity hotspot'*

Sustainable development – challenges and opportunities

There are no 'soft' transition zones or ecological buffers between the preserved wetland and the commercial world beyond. This means that developmental pressures (e.g. intensive agriculture) have a greater impact as there is dramatic land-use change directly adjacent to the reserve – a sharp environmental gradient.

The Wicken Fen Vision Plan

In 1999 the National Trust launched the Wicken Fen Vision Plan, an integrated management plan, which aimed to:

- extend the nature reserve to a maximum of 5300 ha
- restore its habitats to create a 'sustainable landscape-scale space for wildlife and people'.

The aim of the Vision Plan is to create a mosaic of natural wetland habitats. The Trust recognises the need to integrate wildlife conservation with the needs of local people, the economy of the area and tourism. This holistic approach is receiving widespread support from the local community.

Sixty second summary

- Wicken Fen comprises natural habitats (e.g. meadows, fen wetlands and lodes) that support a wide range of biodiversity.
- Agricultural development and urban growth exist on the edges of Wicken Fen, which create management challenges.
- The National Trust's Wicken Fen Vision Plan aims to create sustainable space for wildlife and people.

Over to you

Practice writing a short paragraph explaining how Wicken Fen is meeting the challenges of development on its doorstep.

Glossary

accretion Coastal sediment being deposited on a beach making it wider

active layer The surface layer that is thawed seasonally in periglacial environments

adaptation Fitting into the natural system and local ways of life, which may involve changing behaviour

alluvial fan Triangular fan-shaped alluvial deposit formed at the edge of a mountain front at the outlet of wadis and canyons

aridity index A measure of aridity – it is the ratio between mean annual precipitation (P) and mean annual potential evapotranspiration (PET)

atmosphere The air that surrounds the Earth

bajada Extensive apron of alluvium formed by the coalescence (merging) of alluvial fans

barchan Crescent-shaped sand dune formed at right angles to the prevailing winds

basal sliding Large-scale and often quite sudden movement of a portion of ice in a glacier, usually lubricated by subglacial meltwater

berm A ridge or plateau on the beach formed by the deposition of beach material by wave action

biodiversity The range (diversity) of living organisms within a species, between species and in an ecosystem

biome A large-scale ecosystem (e.g. tundra, savanna)

block separation Rocks with a clear pattern of joints and bedding plains break apart in the form of blocks

blockfield Rock-strewn landscape caused by frost action

burial and compaction Where organic matter becomes buried and is then compressed by the overlying sediment

butte Isolated pillar of horizontally bedded rock, a relic of an eroded mountain landscape

carbon budget A way of using data to describe the amount of carbon that is stored and transferred within the carbon cycle

carbon cycle The recycling of carbon between the main carbon stores – the atmosphere, the lithosphere, the hydrosphere and the biosphere

carbon sequestration An umbrella term used to describe the long-term storage of carbon in plants, soils, rock formations and oceans

carbon sink Anything that absorbs more carbon than it releases

carbon source Anything that releases more carbon than it absorbs

channel flash flooding Flooding that happens after a storm when the water makes its way into channels in the landscape

chemical weathering The breakdown or decay of rocks involving a chemical change

climatic climax vegetation The vegetation that would evolve in a climate region if the seral progression is not interrupted by human activity, tectonic processes etc.

closed system A system with no inputs or outputs

coastal morphology The origin and evolution of a coast

cold-based glacier Glacier where the base temperature is too low to enable liquid water to be present so the glacier freezes to the ground

combustion The process where carbon is burned in the presence of oxygen and converted to energy, carbon dioxide and water

compressional flow Piling up or thickening of glacier ice due to a decrease in the long profile valley floor gradient

constructive wave A powerful wave with a strong swash that surges up a beach usually forming a berm

continentality The influence of a large land mass on weather and climate

cryosphere The frozen parts of the Earth's surface including ice caps, frozen oceans, glaciers and snow cover

cusp Crescent-shaped beach formations with graded sediment; coarse material collects at the 'horns' and finer material collects in the 'bay' area

dalmatian coast A submergent landscape of ridges and valleys running parallel to the coast

decomposition The process where carbon from the bodies of dead organisms is returned to the air as carbon dioxide

deflation The process of wind erosion that involves the removal of loose material from the desert floor, often resulting in the exposure of the underlying bedrock

desert pavement Stony desert surface often resembling a cobbled street

desertification Turning marginal land into a desert by destroying its biological potential

destructive wave A wave formed by a local storm that crashes onto a beach and has a powerful backwash

drift-aligned beach Formed when beach deposits (sand and pebbles) are transferred along a coastline by longshore drift, and accumulate to form a wide beach at a headland where the lateral drift is interrupted

dynamic equilibrium A state of balance where inputs equal outputs in a system that is constantly changing

ecosystem A system in which organisms interact with each other and with their environment

endoreic river River that terminates in a desert region usually in a lake

ephemeral river River that flows intermittently in a desert region

episodic flash flooding Infrequent high intensity rainfall event that results in either sheet flash flooding or channel flash flooding

esker Sinuous (winding) ridge found on the floor of a glacial trough, formed by fluvioglacial deposition in a meandering subglacial river

eustatic change Variations in relative sea level resulting from changes in the amount of liquid water entering the oceans (e.g. glacial meltwater at the end of an ice age)

evapotranspiration the combined losses of moisture through transpiration and evaporation

exfoliation The peeling or flaking of the outer skin of rocks due to intense heating and cooling

exogenous river River that flows continuously through a desert and has its source in mountains outside the desert

extensional flow Stretching or thinning of ice (glacier) in response to an increase in gradient

Glossary

fjord Created when a rise in sea level floods a deep glacial trough

flood hydrograph A graph that plots river discharge against time

flows/transfers The process of moving water or carbon from one store to another

frost heave Small-scale upwards displacement of soil particles as a result from the freezing and expansion of water just below the ground surface

frost shattering Repeated freezing and thawing of water trapped within a rock, causing it to shatter

glacial budget The balance between inputs and outputs of a glacier

global atmospheric circulation system The large-scale circulation of the atmosphere involving three distinct but interconnected circulation cells

granular disintegration The crumbling and breaking down of rocks made of grains (such as granite and sandstone into grains of sand)

gravitational sliding The movement of tectonic plates as a result of gravity

halosere Vegetation succession that originated in an area of saline water

high-energy environments Coastline with powerful waves where rates of erosion exceeds rates of deposition

hydrosere Vegetation succession that originated in an area of fresh water

hydrosphere All of the water on or surrounding the Earth, including oceans, seas, lakes, rivers and the water in the atmosphere

ice wedge V-shaped ice-filled features formed by the enlargement of a surface cracks by frost action. In time the cracks will become infilled with sediment.

inselberg Rounded isolated outcrop of rock, a relic of an eroded upland

insolation The amount of heat (short-wave radiation) that reaches the ground surface

internal deformation Small-scale inter- and intra-granular movement or deformation of ice crystals in response to gravity and mass

isostatic change Rising or falling of a land mass relative to the sea resulting from the release of the weight of ice after the last ice age or by the weight of sediment being deposited

kame Mound or hillock found on the floor of a glacial trough formed by fluvioglacial deposition

lahar Mudflow composed mainly of volcanic ash mixed with water from a crater lake, snowmelt, glacier melt or prolonged torrential rain

liquefaction The jelly-like state of silts and clays resulting from intense ground shaking. This may result in subsidence and collapse of buildings following an earthquake.

lithosere A vegetation succession that originated on a bare rocky surface

lithosphere The outermost solid layer of the Earth, approximately 100 km thick, comprising the crust and upper mantle

low-energy environments Coastline with waves of relatively low power where rates of deposition exceeds rates of erosion

magma plume A rising column of hot rock usually at a plate margin (but can also burn through a plate) creating a hot spot

meltwater channel Often narrow and steep-sided valleys formed by torrents of meltwater at the end of a glacial period

mesa Table-like relic landform formed in horizontally bedded rocks

mitigation Reducing or alleviating the impacts or severity of adverse conditions or events

negative feedback A cyclical sequence of events that damps down or neutralises the effects of a system

nivation Snow-related processes, such as weathering and mass movement, that operate collectively to form shallow hollows in the landscape

nutrient cycle Recycling of nutrients between living organisms and the environment

offshore bars Submerged (or partly exposed) ridges of sand or coarse sediment created by waves offshore from the coast

open system A system with inputs from and outputs to other systems

outwash plain An extensive, gently sloping area of sands and gravels formed in front of a glacier

patterned ground Concentration of large stones on the ground surface, usually associated with polygonal patterns of ice wedges

pediment Gently sloping, usually concave rock surface at the foot of a mountain front

permafrost Permanently frozen soil and rock, a key characteristic of a periglacial environment

photosynthesis The process whereby plants use the light energy from the Sun to produce carbohydrates in the form of glucose

pingo Ice-cored mound formed by the freezing of subsurface water bodies and subsequent swelling of the ground surface

pioneer community The flora and fauna that first colonise a habitat

pioneer species The first plants that colonise an area, usually with special adaptations

plagioclimax community The vegetation succession that results from human influence

playa Salt lake formed on flat clay deposits on a desert plain characterised by high levels of salinity

positive feedback A cyclical sequence of events that amplifies or increases change

psammosere Vegetation succession that originated in a coastal sand dune area

raised beach The result of isostatic recovery which raises wave-cut platforms and their beaches above the present sea level

resilience The psychological quality of strength of character, of being able to respond positively to adversity

respiration A chemical process that happens in all cells, which converts glucose into energy

ria A sheltered winding inlet with irregular shoreline

ridge push The higher elevation at a mid-ocean ridge causes gravity to push the lithosphere that is further from the ridge

runnel The dips in the foreshore area of a beach between ridges. They are drained down the beach by channels that break the ridges.

salt crystallisation Growth of salt crystals within a rock that can cause it to break apart

saltation Rocks or sand that is moved in a series of leaps across a river or sea bed or the desert floor

saltmarsh Coastal ecosystem formed on mudflats (e.g. in a river estuary) largely comprising salt-tolerant plants

sediment budget An attempt to quantify the various stores and transfers associated with sediment movement

sediment cell A conceptual way of describing sediment movement from a source, through various transfers to a sink or output. The movement is usually cyclical.

seif dune Linear sand dune formed parallel to the prevailing winds

seismicity The frequency and distribution of earthquakes in an area

seral stage A stage within the sere

sere A complete vegetation succession

shattering The breakdown of rocks that do not have separate grains or a clear pattern of joints to form angular fragments

sheet flooding Flooding that happens after a storm when water flows across the landscape rather than being diverted into valleys and gullies

slab pull Following subduction, the lithosphere sinks into the mantle under its own weight, helping to 'pull' the rest of the plate with it

soil water Water that is stored in soil

soil water budget The seasonal pattern of water availability for plant growth

stemflow Water flowing down the stems of plants or the trunks of trees

store An accumulation or quantity of water or carbon

subduction Occurs when one tectonic plate slides beneath another, moving down into the mantle. This usually involves oceanic crust sliding beneath continental crust.

submarine volcanoes Volcanoes formed beneath the sea (either a single-vent or fissure volcano), where lava is emitted along a crack in the Earth's crust (e.g. the Mid-Atlantic Ridge)

supervolcano A huge volcano that often takes the form of a caldera and is associated with massive eruptions capable of having a global impact on people

surface creep Sand that is transported by being rolled along the desert floor

suspension Sand that is whisked up and carried by the wind often over great distances. This also applies to sediments being carried along within a water body.

swash-aligned beach A beach formed in a low-energy environments by waves roughly parallel to the shore

symbiotic Diverse organisms that exist in the same environment, often depend on this relationship to survive and prosper

system An approach that usually takes the form of a diagram representing the different components and their interrelationships or links between them

tephra Pyroclastic material that ranges in size from dust to blocks the size of cars

terracettes Steps formed on a slope caused by the freezing and thawing of the ground causing particles to move downhill

thermal fracture A form of weathering brought about by the expansion and contraction of the outer surface of a rock caused by intense temperature fluctuations

thermokarst A typical periglacial landscape of hollows and hummocks

tombolo Ridge of beach material that has formed between an island and the mainland

transform fault A fault that cuts across a mid-ocean ridge

trophic level An organism's position in the food chain

ventifact Sharply angled individual rock usually found on a desert pavement and formed by abrasion

vulcanicity The process of molten rock and gases extruding onto the Earth's surface or intruding into the Earth's crust

wadi Dry river channel, gully or valley formed by periodic water erosion

warm-based glacier Glacier where the base temperature is high enough to enable meltwater to exist and therefore basal sliding to occur

water balance An equation used to express the relationship between precipitation, runoff, evapotranspiration and storage

water cycle The recycling of water between the main water stores – the lithosphere, the hydrosphere, the cryosphere and the atmosphere

weathering The breakdown or decay of rocks in their original place at, or close to, the surface. Chemical weathering involves the absorption of carbon dioxide from the atmosphere.

wildfire An uncontrolled fire, either natural or human-made, that occurs in open country or wilderness

yardang Ridge formed by abrasion of vertical bands of resistant rock

zeugen Ridge formed in horizontal rocks, with a clear resistant cap rock

Revision planner

Date: _____

	Revision Period 1	Revision Period 2	Revision Period 3	Revision Period 4	Revision Period 5
Monday					
Tuesday					
Wednesday					
Thursday					
Friday					
Saturday					
Sunday					

Date: _____

	Revision Period 1	Revision Period 2	Revision Period 3	Revision Period 4	Revision Period 5
Sunday					
Saturday					
Friday					
Thursday					
Wednesday					
Tuesday					
Monday					